Fortschritte der Chemie organischer Naturstoffe

Progress in the Chemistry of Organic Natural Products

55

Founded by L. Zechmeister
Edited by W. Herz, H. Grisebach, G.W. Kirby, and Ch. Tamm

Authors:
M. T. Davies-Coleman, J. Galambos, L. Hough,
C. E. James, R. Khan, K. Krohn,
M. Lounasmaa, D. E. A. Rivett

Springer-Verlag
Wien New York 1989

Dr. W. Herz, Professor of Chemistry, Department of Chemistry,
The Florida State University, Tallahassee, Florida, U.S.A.

Prof. Dr. H. Grisebach, Biologisches Institut II, Lehrstuhl für Biochemie der Pflanzen,
Albert-Ludwigs-Universität, Freiburg i.Br., Federal Republic of Germany

G.W. Kirby, Sc. D., Regius Professor of Chemistry, Chemistry Department,
The University, Glasgow, Scotland

Prof. Dr. Ch. Tamm, Institut für Organische Chemie der Universität Basel,
Basel, Switzerland

This work is subject to copyright.
All rights are reserved, whether the whole or part of the material is concerned, specifically those of translation, reprinting, re-use of illustrations, broadcasting, reproduction by photocopying machine or similar means, and storage in data banks.

© 1989 by Springer-Verlag/Wien

Library of Congress Catalog Card Number AC 39-1015

Printed in Austria

ISSN 0071-7886

ISBN 3-211-82087-6 Springer-Verlag Wien-New York
ISBN 0-387-82087-6 Springer-Verlag New York-Wien

Appearance of the present volume 55 of "Progress in the Chemistry of Organic Natural Products" marks the 100th anniversary of the birth, on May 14, 1889, of Professor László Zechmeister, the founder of our series. First in Pécs, Hungary, then, after a hiatus caused by World War II, in Pasadena, California, he edited 27 volumes of what came to be known simply as "Zechmeister" until he retired at the age of 80 in 1969. Those who knew him remember him for his contributions to science, for the stimulus he provided for his contemporaries and his students and, in his capacity as editor, for his meticulous attention to detail.

The Editors

Contents

List of Contributors	X
Naturally Occurring 6-substituted 5,6-dihydro-α-pyrones. By M.T. DAVIES-COLEMAN and D.E.A. RIVETT	1
1. Introduction	1
2. Nomenclature and Classification	2
3. Distribution and Structure Determination	3
3.1. 6-Alkyl-5,6-dihydro-α-pyrones	4
3.2. 6-Alkenyl-5,6-dihydro-α-pyrones	10
3.3. 6-Aryl-5,6-dihydro-α-pyrones	16
4. Physical Methods of Structure Determination	19
4.1. ^{1}H- and ^{13}C-NMR Spectroscopy	20
4.2. Mass Spectrometry	21
4.3. Infrared and Ultraviolet Spectroscopy	23
4.4. Circular Dichroism	23
5. Biosynthesis	26
6. Biological Activity	28
References	29
Addendum	35
Building Blocks for the Total Synthesis of Anthracyclinones. By K. KROHN	37
I. Introduction	37
II. General Synthetic Methods	39
1. Friedel-Crafts Reactions	39
2. Diels-Alder Reactions	42
3. Anionic Reactions	45
4. Transition Metal Mediated Reactions	48
III. Synthesis of Building Blocks for Rings A and B	49
IV. Construction of Ring A	65
1. Diels-Alder Reactions	65
2. Electrophilic Additions	67
3. Nucleophilic Additions	70
V. Building Blocks for Rings C and D	72
VI. Concluding Remarks	76
References	77

Indole Alkaloid Production in *Catharanthus roseus* Cell Suspension Cultures. By M. LOUNASMAA and J. GALAMBOS 89

1. Introduction 89
2. Discussion 90
   2.1. Indole Alkaloids and Their Formation in the Plant and in Cell Cultures . . . 90
   2.2. General Methods of *Catharanthus roseus* Cell Suspension Culture Work . . . 97
      2.2.1. Analytical Methods 97
      2.2.2. Development of High Yielding Cell Lines 98
      2.2.3. Growth and Alkaloid Production in *Catharanthus roseus* Cell Suspension Cultures 99
      2.2.4. Effects of Culture Conditions on Growth and Alkaloid Production . . 101
         2.2.4.1. Medium Composition 101
         2.2.4.2. Addition of Precursors 103
         2.2.4.3. Light 104
         2.2.4.4. Temperature 104
         2.2.4.5. Gaseous Environment 105
      2.2.5. One-stage Systems 105
      2.2.6. Large-scale Fermentation 105
      2.2.7. Immobilized Cell Systems 106
      2.2.8. Cell Free Systems 106
3. Conclusions 107
References 108

Sucrose and Its Derivatives. By C.E. JAMES, L. HOUGH, and R. KHAN 117

1. Introduction 118
   1.1. History 118
   1.2. Structure 119
   1.3. Nomenclature 121
   1.4. Synthesis 122
   1.5. Biosynthesis 125
   1.6. Conformation 127
2. Protective and Functional Groups 129
   2.1. Ethers 129
      2.1.1. Tritylation 129
      2.1.2. Methylation 130
      2.1.3. Silylation 131
   2.2. Cyclic Acetals 132
   2.3. Esters 134
      2.3.1. Carboxylates 135
         2.3.1.1. Acetates 135
         2.3.1.2. Benzoates 138
         2.3.1.3. Benzoylpropionates 139
         2.3.1.4. Pivalates 139
      2.3.2. Sulphonates 140
      2.3.3. Chlorosulphates 143
      2.3.4. Other Esters 148
3. Derivatives 149
   3.1. Anhydrides (Oxetanes) and Epoxides (Oxiranes) 149
   3.2. Halides 154
   3.3. Unsaturated, Deoxy and Branched-Chain Compounds 157

   3.4. Nitrogen-containing Compounds: Azides, Amines and Morpholines 159
   3.5. Sulphur Derivatives 163
4. Enhancement of Sweetness: Structure Activity Relationships 165
5. Natural Products Containing Sucrose 168
   5.1. β-D-Fructofuranosyl Derivatives 168
   5.2. α-D-Glucopyranosyl Derivatives 169
   5.3. α- and β-D-Galactopyranosyl Derivatives 170
   5.4. Galloyl Derivatives 171
   5.5 Agricinopine A 172
   5.6. Sucrose Esters from Potato 173
   5.7. Sucrose Esters from Tobacco 173
Acknowledgment 175
References 175
Author Index 185
Subject Index 197

List of Contributors

DAVIES-COLEMAN, Dr. M.T., Department of Chemistry and Biochemistry, Rhodes University, P.O. Box 94, Grahamstown, 6140, Republic of South Africa.

GALAMBOS, Dr. J., Laboratory for Organic and Bioorganic Chemistry, Technical University of Helsinki, SF-02150 Espoo 15, Finland.

HOUGH, Prof. L., Department of Chemistry, King's College London, University of London, Kensington Campus, London W8 7AH, U.K.

JAMES, Dr. C.E., Department of Chemistry, King's College London, University of London, Kensington Campus, London W8 7AH, U.K.

KHAN, Dr. R., Tate and Lyle Research and Technology, Philip Lyle Memorial Research Laboratory, P.O. Box 68, Whiteknights, Reading, Berkshire RG6 2BX, U.K.

KROHN, Prof. K., Institut für Organische Chemie der Technischen Universität, Hagenring 30, D-3300 Braunschweig, Federal Republic of Germany.

LOUNASMAA, Prof. Dr. M., Laboratory for Organic and Bioorganic Chemistry, Technical University of Helsinki, SF-02150 Espoo 15, Finland.

RIVETT, Prof. D.E.A., Department of Chemistry and Biochemistry, Rhodes University, P.O. Box 94, Grahamstown, 6140, Republic of South Africa.

Naturally Occurring 6-substituted 5,6-dihydro-α-pyrones

By M.T. DAVIES-COLEMAN and D.E.A. RIVETT, Department of Chemistry and Biochemistry, Rhodes University, Grahamstown, Republic of South Africa

Contents

1. Introduction 1
2. Nomenclature and Classification 2
3. Distribution and Structure Determination 3
   3.1. 6-Alkyl-5,6-dihydro-α-pyrones 4
   3.2. 6-Alkenyl-5,6-dihydro-α-pyrones 10
   3.3. 6-Aryl-5,6-dihydro-α-pyrones 16
4. Physical Methods of Structure Determination 19
   4.1. ^{1}H- and ^{13}C-NMR Spectroscopy 20
   4.2. Mass Spectrometry 21
   4.3. Infrared and Ultraviolet Spectroscopy 23
   4.4. Circular Dichroism 23
5. Biosynthesis 26
6. Biological Activity 28
References 29
Addendum 35

1. Introduction

Naturally occurring 5,6-dihydro-α-pyrones have up to now not been reviewed *per se* but have been included in the reviews of α-pyrones [1,2]. Since approximately 65 of these compounds are presently known a review devoted entirely to them is now desirable. Moreover, the newest review [2] is not comprehensive, deals only slightly with their stereochemistry and does not discuss the assignment of absolute configuration by techniques such as circular dichroism.

6-Substituted 5,6-dihydro-α- or 2-pyrones possess a diverse range of biological activity. They have been reported as plant growth inhibitors (*3*), insect antifeedants (*4*), antifungal (*5*) and antitumour (*6*) agents. They are widely distributed in both plants and fungi and have been isolated from thirteen families of plants and twenty fungal species. α-Pyrones and 5,6-dihydro-α-pyrones rarely occur together and have been found in admixture in only an unidentified *Penicillium* species (*7*), *Alternaria citri* (*8*) and *Piper methysticum* (*9*).

This review covers the literature included in Chemical Abstracts up to December 1987 and excludes compounds where a steroid nucleus is attached at C-6 to the lactone ring, e.g. withanolides. The main emphasis is on distribution, structure elucidation and absolute stereochemistry. Interesting aspects of biosynthesis, biological activity and pharmacology are also discussed but synthetic methods are not except where used to confirm structure and stereochemistry.

2. Nomenclature and Classification

There is little uniformity within the nomenclature of 5,6-dihydro-α-pyrones. According to IUPAC recommendations, lactones can be named in either one of two ways (*10*). The first or classical method identifies the lactone with a corresponding heterocyclic nucleus. Therefore, a substituted αβ-unsaturated-δ-lactone is named as a substituted 5,6-dihydro-2H-pyran-2-one, e.g. 6-ethyl-5,6-dihydro-2H-pyran-2-one (**1**). This is the most common procedure and is that accepted by Chemical Abstracts. Where no ambiguity exists, it is permissible to omit the indicated hydrogen, H, its position and also to reduce pyran-2-one to 2-pyrone. The term α-pyrone is synonomous with 2-pyrone and in accordance with common usage will be employed here.

The lactonisation of a carboxylic acid hydroxylated at C-5 and replacement of the suffix 'oic' with 'olide' forms the basis of the second recommended nomenclature system, e.g. 2-hepten-5-olide or 5-ethyl-2-penten-5-olide (**2**) (*10*).

(**1**)   (**2**)

The numbering of the δ-lactone ring differs in (**1**) and (**2**) and this can lead to confusion when specifying ring positions. To minimise uncertainty the numbering sequence of the former nomenclature system

References, pp. 29–35

will be used here. Ring and side chain positions are differentiated by numbering the C-6 side chain with prime numbers.

The trivial names of the 5,6-dihydro-α-pyrones under review have been retained. Where trivial names have not been given to individual compounds, names in accordance with the Chemical Abstracts nomenclature system are used.

In the review of MORS *et al.* (*1*) methoxylation at C-4 was used as the criterion for the classification of α-pyrones because only two compounds, phenylcoumalin (**3**) and paracotin (**4**), had been found in early studies of aromatic α-pyrones not to contain a C-4 methoxy group. This structural deficit was related to a more recent evolutionary origin of the plant species containing these compounds within their respective genera. MORS *et al.* (*1*) were able to support their argument with morphological evidence. However, since a C-4 methoxy group occurs in only thirteen of the *ca.* 65 reported 6-substituted 5,6-dihydro-α-pyrones and such substitution is thus the exception rather than the rule, this classification method will not be used here for purposes of subdivision. Instead, compounds are discussed in order of increasing complexity of substitution around the lactone ring.

In their review ADITYACHAUDHURY and DAS (*2*) have divided the α-pyrones and 5,6-dihydro-α-pyrones into groups based on the substituent at C-6. Although this classification does not have any recognisable biosynthetic foundation, it is retained here because of the order it imposes on an otherwise structurally diverse group of compounds. They restricted their division to 6-aryl- and 6-alkyl-α-pyrones and -5,6-dihydro-α-pyrones. Since nearly a quarter of the 5,6-dihydro-α-pyrones reported have a 6-alkenyl side chain a third group has been introduced. Compounds with a 6-styryl-substituent are classified with 6-aryl-5,6-dihydro-α-pyrones.

3. Distribution and Structure Determination

6-Substituted 5,6-dihydro-α-pyrones are widely distributed in the plant kingdom but are particulary associated with the Lamiaceae, Piperaceae, Lauraceae and Annonaceae. They have been found in all parts of plants including the leaves, stems, flowers and fruit and are also widespread amongst fungi. Early structural work involved the use of a number of chemical degradation methods such as periodate oxidation, ozonolysis, oxidation and hydrolysis but recent advances in spectroscopic techniques have considerably facilitated structure determinations. However, the assignment of absolute stereochemistry with some degree of certainty is more difficult. The high level of oxygenation in the acyclic substituents of some compounds often results in a complex sequence of adjacent chiral centres. A wide diversity of methods, both spectroscopic and chemical, has been used to ascertain the absolute stereochemistry of these compounds and is reviewed here.

3.1. 6-Alkyl-5,6-dihydro-α-pyrones

The simplest of these compounds is parasorbic acid (**5**). This volatile compound was obtained by HOFMANN in 1859 from steam distillation of an acidified extract of mountain ash berries, *Sorbus aucuparia* (Rosaceae) (*11*). The action of alkali on the oily (**5**) yielded an amorphous salt, later shown to be the salt of an unsaturated hydroxy acid, $C_6H_{10}O_3$ (*12*). Although (**5**) was one of the first αβ-unsaturated-δ-lactones to be isolated, it took nearly a century before KUHN and JERCHEL (*13*) established its 6-methyl-5,6-dihydro-2H-pyran-2-one structure.

(5)   (6)

Almost two decades later two independent groups established the (6*S*)-configuration of (**5**) by degradation to (+)-(*S*)-hexan-1,5-diol (*14*) and to (+)-(3*S*)-hydroxybutyric acid (*15*). Although a number of syntheses of racemic (**5**) have been reported the first synthesis of the naturally occurring (+)-(*S*)-enantiomer from 3,4-di-O-acetyl-L-rhamnal was only recently described by LICHTENTHALER *et al.* (*16*) who confirmed the earlier stereochemical assignment. TSCHESCHE *et al.* (*17*) showed that parasorbic acid is not present in the plant as (**5**) but as parasorbiside (**6**).

The short chain alkyl homologues of (**5**) are also volatile, and are important aroma components in both food and beverages. Most of

References, pp. 29–35

these compounds have a low odour threshold (*ca.* 0.1 ppm) and possess significant flavour value. The distribution and odour descriptions of both saturated and unsaturated δ-lactones have been reviewed (*18*).

6-Propyl-5,6-dihydro-α-pyrone (**7**) was detected by GC-MS as a neutral aroma constituent of burley tobacco, *Nicotiana tabacum* (Solanaceae) (*19*). Massoilactone (**8**) was first found in the bark oil of *Cryptocarya massoia* (Lauraceae), a New Guinean medicinal plant (*20*). The structure proposed by MEIJR was later confirmed by ABE and SATO (*21*) through isolation and identification of the acids obtained from oxidative degradation. The (*R*)-configuration of (**8**) assigned to the naturally occurring (−)-enantiomer by comparison of its ORD curve with that of (**5**) was confirmed by MORI (*22*) who synthesised the (+)-(6*S*)-enantiomer from (+)-(*R*)-glyceraldehyde. PIRKLE and ADAMS (*23*) have synthesised both the (−)-(*R*)- and (+)-(*S*)-enantiomers of massoilactone by a different route. Their key step involves the separation of disastereomeric derivatives of the racemic lactone precursors by HPLC, and thus avoids the partial racemisation encountered by MORI.

Massoilactone has also been obtained from tuberose flowers (*24*), wine (*25*), molasses (*26*) and two species of Australian ants where it forms part of the ants' defence mechanism (*27*). A similar compound, 6-heptyl-5,6-dihydro-α-pyrone (**9**), is a minor volatile constituent of *C. massoia* (*27*).

Several compounds oxygenated in the side-chain or lactone ring are known. BOHLMANN and SUWITA (*28*) proposed structure (**10**) but without any stereochemistry for tarchonatus lactone isolated from *Tarchonatus trilobus* (Compositae). The absolute configuration has recently been proven by synthesis (*29*).

A (6*R*)-configuration was proposed for boronolide (**11**), first isolated by FRANCA and POLONSKY (*30*) from *Tetradenia fruticosa* (Lamiaceae), by application of Hudson's lactone rule to the molecular rotation. This result was used in conjunction with a recent crystal structure analysis (*31*) to assign the absolute stereochemistry shown. Chemical evidence for the (6*R*)-configuration of (**11**), isolated from *T. barberae*, has been obtained by oxidative degradation to D-malic acid, recognised by GC-MS as the acetylated 2(+)-dibutyl ester (*32*). This simple method of determining the configuration at C-6 deserves wider application in related compounds.

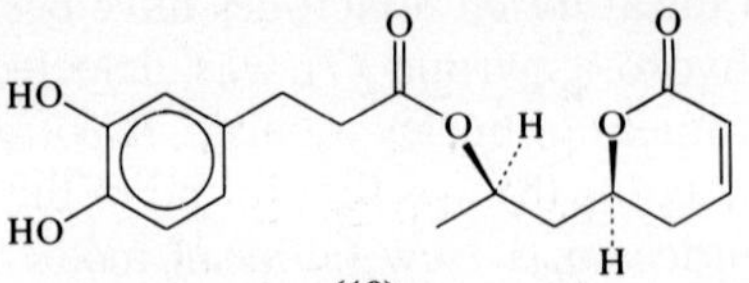

(10)

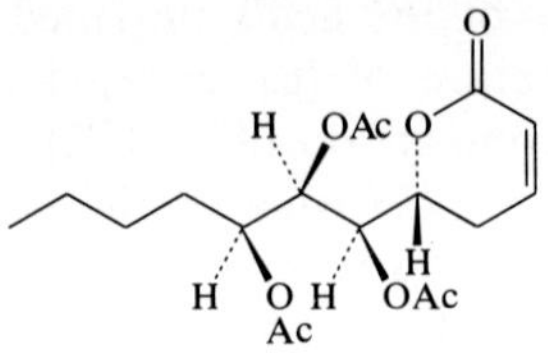

(11)

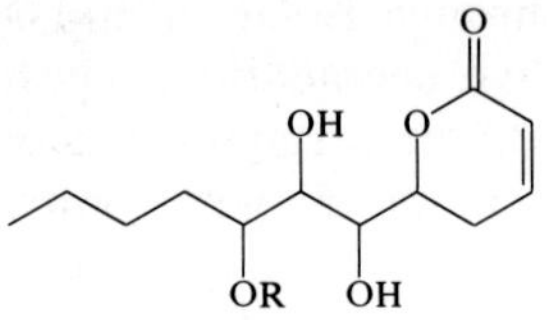

(12) R=H
(13) R=Ac

Two related deacetyl derivatives of (11), deacetylboronolide (12) and 1,2-dideacetylboronolide (13) have been isolated by VAN PUYVELDE et al. (33, 34) from another species, *T. riparia* (formerly *Iboza*), which is a common medicinal plant in central Africa. The structures of (12) and (13) were proposed from ^{1}H-NMR and MS data. Paradoxically, (12) was reported to be optically inactive and GC-MS results suggested that it was a mixture of three stereoisomers. No optical rotation has been reported for (13) and the stereochemistry remains unassigned.

Two unusual C_{26}-polyhydroxy lactones (14) and (15) have been isolated from *Eupatorium pilosum* Walt., a North American species of the Compositae (35). The structures of these compounds but without the stereochemistry in the side-chain were determined with the aid of ^{1}H-NMR and MS. Their (6S)-configuration, assigned from the negative sign for the CD peak, is incorrectly described as (6R). NAKATA et al. (29) proposed an all *syn* stereochemistry for triol (14) and then proceeded to synthesise the racemic acetate of this compound which had an NMR spectrum identical with that of the triacetate of (14). By analogy the tetraol (15) probably also has the (2'S, 4'R, 6'R, 6S) stereochemistry but with the configuration at C-10' unknown but probably (R) (36).

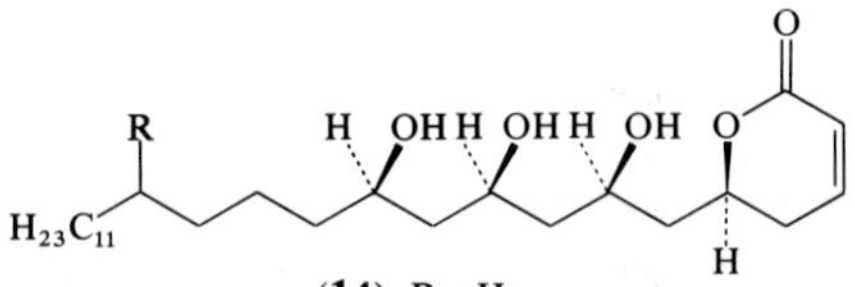

(14) R=H
(15) R=OH

References, pp. 29–35

A number of fungal metabolites have been reported which have one or more substituents at other positions around the lactone ring. Three of these compounds are methoxylated at C-4. The structure of pestalotin (16) a gibberellin synergist first isolated by KIMURA et al. (37) from the culture filtrate of a phytopathogenic fungus, *Pestalotia cryptomeriaecola*, was established by means of standard physical techniques. The proposed (6R)-configuration results from an incorrect interpretation of Snatzke's rules relating to the negative sign of the Cotton effect in the CD curve of (16). ELLESTAD, MCGAHREN and KUNSTMANN (38) isolated (16) shortly afterwards from an unidentified *Penicillium* species and named it LL-P880α. They did not acknowledge the earlier work of KIMURA et al. who failed to report the optical rotation, but the melting point of their substance and the negative sign of the Cotton effect at 243 nm of LL-P880α would suggest that it is identical with pestalotin. The (S)-configuration at C-6 assigned from the CD data is supported by results obtained by application of the Horeau method to the methylated base hydrolysis product (17) of LL-P880α. The same method was used to assign a (1'S)-configuration to (16).

(16)   (17)

The first synthesis of naturally occurring (−)-pestalotin was carried out by SEEBACH and MEYER (39) who confirmed the stereochemical assignments of ELLESTAD et al. This asymmetric synthesis yielded both (−)-(16) and (−)-epipestalotin which were separated by repeated crystallisation. The two contiguous chiral centres of (16) have been an attractive target for synthesis. Further syntheses of (−)-(16) have been reported by ICHIMOTO et al. (40) from D-glucose and by MASAKI et al. (41) from (+)-(R,R)-diethyl tartrate. All four possible stereoisomers of (16) have recently been synthesised (42).

ELLESTAD et al. (7) later obtained two further metabolites, LL-P880β (18) and its unsaturated analogue LL-P880γ (19), from the same unidentified *Penicillium* species. The (1'S)- and (2'R)-configuration of the vicinal diol in (18) were determined by the excitation chirality CD method.

STRUNZ et al. (43) isolated (16) and 6-[1'-oxo-pentyl]-5,6-dihydro-α-pyrone (20) from an unidentified fungus and showed that the latter compound was identical with the Jones oxidation product of (16).

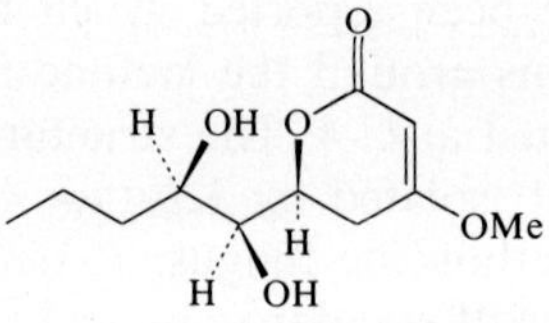

(18)

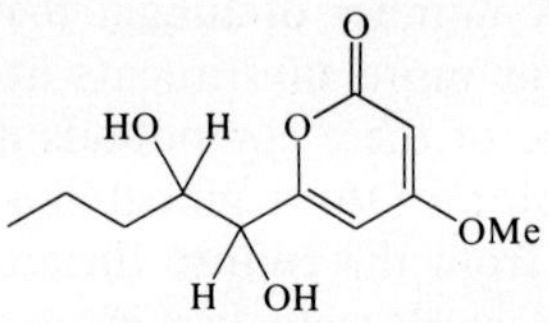

(19)

Alkylation at C-6 and oxygenation at C-5 are structural features found in a number of 5,6-dihydro-α-pyrones isolated from both plants and fungi. A high incidence of possible dietary related gastric tumours in Japan prompted an investigation of the edible ferns *Osmunda japonica* Thunberg and *Osmunda regalis* var. *spectabilis* Willd (*44*). The major constituent of the ferns was shown to be osmundalin (**21**) which gave osmundalactone (**22**) on acid hydrolysis. Recently (**22**) was isolated from *O. japonica* and shown to possess strong insect anti-feedant activity (*4*). The structure of (**22**) was proposed through a variety of spectroscopic and chemical methods. Reduction of dihydro-osmundalactone with lithium aluminium hydride (LAH) gave 1,4,5-hexanetriol which on periodate oxidation yielded acetaldehyde and 4-hydroxybutanal and thus established the position of the C-5 hydroxyl group. The configurations of the chiral centres at C-5 and C-6 were inferred from CD measurements and confirmed by synthesis of (**22**) from L-rhamnose (*44*).

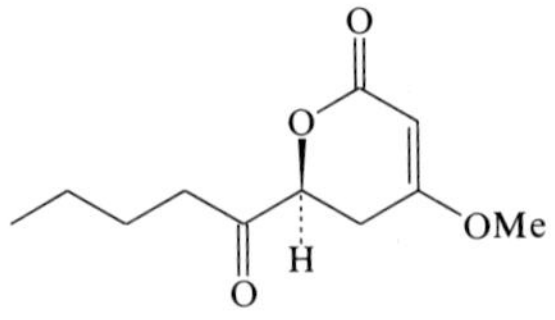

(20)

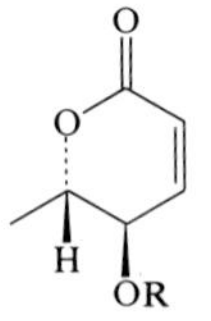

(21) R = β-D-glucosyl
(22) R = H

ARGOUDELIS and ZIESERL (*45*) extracted a compound from the fungus *Aspergillus nidulans* which they named antibiotic U-13,933 (**23**). This compound later became known as asperline and has also been isolated from *A. caespitosus* (*46*), *A. carneus* (*47*) and *Aspergillus* species NRRL 5769 (*48*). The structure and relative stereochemistry was established from extensive ^{1}H-NMR decoupling experiments. EVANS, ELLESTAD and KUNSTMANN (*49*) directly related (**23**) to phomalactone (**24**) from an unidentified *Nigrospora* species by oxidation of the acetylated (**24**) with m-chloroperbenzoic acid to (**23**). Phomalactone has also been isolated from *Phoma minispora* (*50*) and *Nigrospora* species Z 1276

(*51*). The (*S*)-configuration at the two chiral centres in (**24**) is based on CD data (*49*).

(**23**) (**24**) (**25**)

Aspergillus elegans (*47*), *A. ochraceus* (*52*), *A. melleus* (*53*) and an unidentified penicillic acid producing *Aspergillus* species (*54*) have yielded an unusual 5,6-dihydro-α-pyrone, aspyrone (**25**), related to (**23**) and (**24**). The absolute stereochemistry is based on ozonolytic degradation of (**25**) to two chiral fragments, (−)-1-deoxyerythritol and (−)-2,3-epoxybutyric acid (*53*). The absolute stereochemistry of the former fragment was established by comparing the optical rotation of the *tris*-(*p*-nitrobenzoate) derivative with the corresponding ester of (−)-(2*R*,3*S*)-1-deoxyerythritol, and this confirmed the (5*S*,6*R*)-stereochemistry of (**25**) originally proposed from ORD data. The epoxy acid formed a brucine salt identical with that of (−)-(2*R*,3*S*)-epoxybutyric acid, and hence a (1″*S*,2″*S*)-epoxypropyl side chain (confirmed by X-ray structure analysis) is present.

Alkylation at C-6, hydroxylation at C-4 and acylation at C-3 of the lactone ring is found in a number of compounds possessing strong antifungal activity. The first compound of this type to be isolated was alternaric acid (**26**) from the fungus *Alternaria solani* (*5*). The structure of the acyl side chain at C-3 was established by GROVE (*55*) and BARTELS-KEITH (*56*) from a series of chemical degradations which included alkaline hydrolysis, ozonolysis and periodate oxidation. The absolute stereochemistry at C-6 and of the three chiral centres of the acyl substituent were not assigned.

(**26**) (**27**)

Another compound of this general type was recently obtained from the fungus *Lachnellula fuscosanguinea* (*57*). The structure of lachnelluloic acid (**27**), proposed from ^{1}H-NMR and MS data, was confirmed by synthesis of the racemate from the symmetrical diketone, 6,8-tridecanedione.

Podoblastins A, B, C, (**28–30**) are antifungal compounds related to (**26**) and (**27**) and are found in the higher plant *Podophyllum peltatum* L. (Berberidaceae) (*58*). Their chemical structures were deduced from GC-MS of the methylated peroxide oxidation products. The (6*R*)-configuration in all three compounds was established using the Pirkle's method (*59*) in which the carbamate diastereomer of the methylated degradation product (**31**) was shown by HPLC to be identical with N-(*R*)-1-naphthylethyl-O-(*R*)-2-(1-methoxycarbonyl)-pentyl carbamate (**32**).

(**28**) R=CH$_3$(CH$_2$)$_{10}$—
(**29**) R=H$_2$C=CH—(CH$_2$)$_9$—
(**30**) R=CH$_3$(CH$_2$)$_{12}$—

(**31**)

(**32**) R=α-naphthyl

3.2. 6-Alkenyl-5,6-dihydro-α-pyrones

The short chain 6-alkenyl-5,6-dihydro-α-pyrones are also volatile and hence resemble their partially saturated analogues. The first decadienolide found in nature was tuberolactone (**33**) from the absolute of tuberose flowers *Polianthes tuberosa* L. (Amaryllidaceae) (*24*). The structure of (**33**) is based on ^{1}H-NMR and MS studies. The (R)-configuration at C-6 was determined by hydrogenation of (**33**) to (+)-5-decanolide (**34**), previously prepared by microbiological reduction of the corresponding keto-acid (*60*).

(**33**)

(**34**)

Another volatile oil, argentilactone (**35**), has been extracted from the rhizomes of *Aristolochia argentina* Gris. (*Aristolochiaceae*) by PRIESTAP *et al.* (*61*). Hydrogenation of (**35**) followed by saponification of the lactone ring gave 5-hydroxydodecanoic acid. Lanthanide shift reagents were used to assign all the proton signals in the ^{1}H-NMR spec-

References, pp. 29–35

trum of (**35**). The *cis* configuration of the exocyclic double bond and the *pseudo*-equatorial orientation of the C-6 side chain were also determined and used, together with the positive sign of the CD curve, to assign a (R)-configuration to the single chiral centre at C-6. The structure of naturally occurring (−)-(**35**) has been confirmed by synthesis from the chiral aldehyde (**36**) by a Wittig reaction (*62*).

(**35**)   (**36**)

As encountered earlier, oxygenation of the side chain or of the lactone ring reduces the volatility of 6-alkenyl-5,6-dihydro-α-pyrones. The Lamiaceae have yielded seven compounds with varying degrees of oxygenation of the alkenyl side chain. The simplest of these are umuravumbolide (**37**) and deacetylumuravumbolide (**38**) from *Tetradenia riparia* (*33*) whose structures were determined by use of spectroscopic techniques. The *trans* configuration of the exocyclic double bond was proposed from the strong IR absorbance at 965 cm^{-1}. Surprisingly, these compounds are reported as showing no optical rotation. In accordance with the reputed presence of (**37**) and (**38**) as racemates in *T. riparia*, ACHENBACH and WITZKE (*63*) have reported a synthesis of racemic (**37**). A separation of the (6R,3'S)-diastereomer and its enantiomer (6S,3'R) from the racemic mixture by silica gel chromatography is described.

(**37**) R = Ac
(**38**) R = H

The first 6-substituted 5,6-dihydro-α-pyrone to be isolated from the Lamiaceae was hyptolide (**39**) from *Hyptis pectinata* Poit. GORTER (*64*) assigned structure (**40**) to hyptolide by identifying of the acids obtained from hydrogenation and silver oxide oxidation but BIRCH and BUTLER (*65*) found that this structure was at variance with the ^{1}H-NMR spectrum of hyptolide and proposed structure (**41**) from chemical degradation experiments. Hydrogenation of (**41**) required 3.2 mole equivalent of hydrogen. Hydrolysis of the hydrogenation products yielded dextrorotatory 10,11-dihydroxydodecanoic acid and 8,10,11-trihydroxydodecanoic acid, produced by hydrogenolysis of the lactone and the 3'-acetoxy group. The position of the exocyclic double bond at C-1' was thus established and its *trans* configuration followed from the IR absorbance at 965 cm^{-1}.

(**39**)   (**40**)   (**41**)

The absolute stereochemistry of hyptolide (**39**) was recently determined by X-ray analysis which also showed that the configuration of the exocyclic double bond is *cis* and not *trans* (*66*). The (6*R*)-configuration is based on the positive Cotton effect shown by (**39**). The absolute stereochemistry at C-5' and C-6', and hence of the whole molecule, has been conclusively settled by KJAER (*66*) who synthesised the dextrorotatory (10*S*,11*R*)-dihydroxydodecanoic acid obtained by BIRCH and BUTLER.

Hyptis species have yielded a further three 6-alkenyl-5,6-dihydro-α-pyrones which have the same general structure but different stereochemistry at the acyclic chiral centres. Two compounds, anamarine (**42**) (*67*) and olguine (**43**) (*68*) were extracted from an unidentified *Hyptis* species, whilst the third compound, 5-deacetoxy-5'-epi-olguine (**44**) (*69*) was obtained from a common Mexican plant *Hyptis oblongifolia* Bentham. The structures (**42**), (**43**) and (**44**) are based on extensive ^{1}H-NMR decoupling experiments. The *trans* configuration of the exocyclic double bonds, the *cis* configuration of the epoxide groups in (**43**) and (**44**) and the relative stereochemistry of all three compounds were determined through X-ray crystallography. The positive sign for the CD peak of (**42**) established the (*R*)-configuration at C-6 and hence the absolute stereochemistry of the molecule could be assigned. The anomalous X-ray dispersion effect of the oxygen atoms was used to determine the absolute stereochemistry of (**43**). Compound (**44**) is very similar to (**43**) and differs only in the absence of the acetoxy group at C-5 and the reverse configuration at C-5'. The *pref* relationship of C-5' and C-6' in (**44**) was assigned from X-ray and ^{1}H-NMR studies, and hence (5'*R*)- and a (6'*S*)-configurations are present at these chiral centres.

References, pp. 29–35

Retrosynthetic analysis suggests that synthesis of (**42**) could be achieved via a Wittig reaction between aldehyde (**36**) and a bromotetraacetate derived from D-gulonolactone (*70*). VALVERDE *et al.* (*71*) have recently described the synthesis of the natural (−)-isomer (**42**) from D-glucose by a different route.

Synrotolide (**45**), isolated from *Syncolostemon rotundifolius* (*72*), is closely related to compounds (**37**), (**38**), (**39**), (**42**), (**43**), and (**44**). Extensive ^{1}H- and ^{13}C-NMR decoupling experiments supported the gross structure while the relative stereochemistry at the chiral centres and the *cis* nature of the double bond is based on X-ray crystallography. Synrotolide was split by reductive ozonolysis to 6-deoxy-L-allose, identified by GC as the acetylated 2(−)-octyl glycoside and hence the absolute stereochemistry is established.

An unusual compound (5′Z, 8′Z, 11′Z)-6-(heptadeca-5′,8′,11′-trienyl-1′-yl)-5,6-dihydro-2H-pyran-2-one (**46**) has been extracted as a colourless oil from the red alga *Phacelocarpus labillardieri* (Sphaerococaeceae) and the structure determined by ^{1}H- and ^{13}C-NMR *(73)*. The position of the homoallylic triene system in the C_{17} side chain is based on the isolation of hexanal from ozonolysis of **(46)**. The stereochemistry at C-6 remains unassigned.

Three novel phosphate-containing antitumour agents, CI-920 **(47)**, PD 113270 **(48)** and PD 113271 **(49)** have been isolated from a subspecies of *Streptomyces pulveraceus* *(6)*. The chemical structures of these compounds were proposed from extensive ^{1}H-NMR analysis and confirmed by chemical degradation *(74)*. The nature of the side chain was determined by treatment of hydrogenated **(47)** with phosphorus and hydroiodic acid, reduction of the resulting mixture of iodocompounds with LAH, and catalytic hydrogenation to 8-methyl-1-octadecanol. Further evidence for the chemical structure was obtained from periodate oxidation of dephosphorylated **(47)** to the keto-lactone **(50)**, shown by synthesis to possess a (6*R*)-configuration. The stereochemistry at C-3', C-4' and C-6' was not assigned. Another novel antitumour antibiotic kazusamycin B **(51)** has been obtained from *Streptomyces* species No. 81–484 *(75)*.

(47) R_1=H, R_2=OH
(48) R_1=R_2=H
(49) R_1=R_2=OH

(50)

(51)

Six compounds, all derived from geranylnerol, have been reported by BOHLMANN *et al.* *(76)* from *Ichthyothere ulei* Thumb. (Compositae). The structures of **(52)** and ichthyouleolide **(53)** are based on ^{1}H-NMR data. The chemical shifts of the protons in the ^{1}H-NMR spectra of **(54)** and **(56)** revealed the presence of a hydroperoxide group in the C-2 side chain. Addition of triphenylphosphine to solutions of **(54)** and **(56)** afforded the corresponding diols **(55)** and **(57)** which were also present in the plant. The configuration at C-6 remains unassigned in these compounds which are clearly related to acanthoaustralide **(58)** from *Acanthospermum australe* (Compositae) *(77)*.

References, pp. 29–35

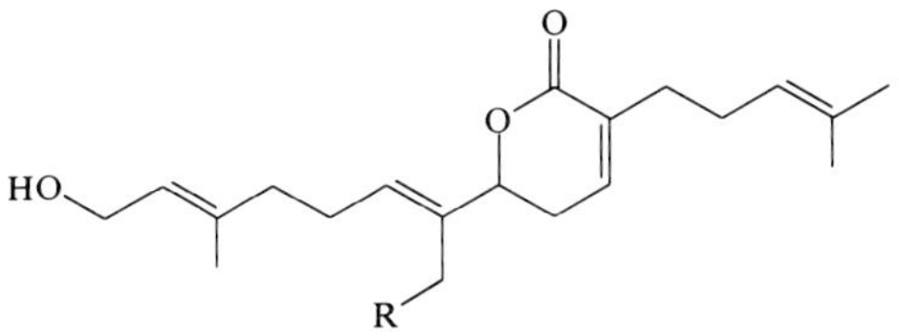

(52) R=H
(53) R=OAc

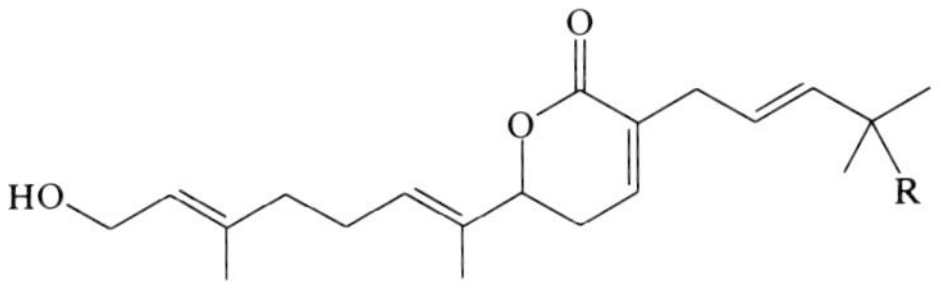

(54) R=OOH
(55) R=OH

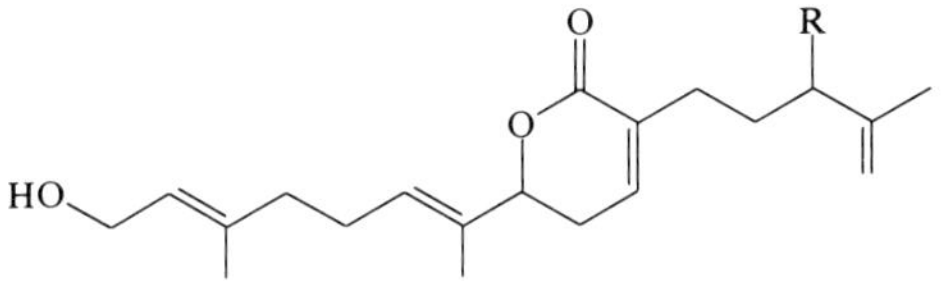

(56) R=OOH
(57) R=OH

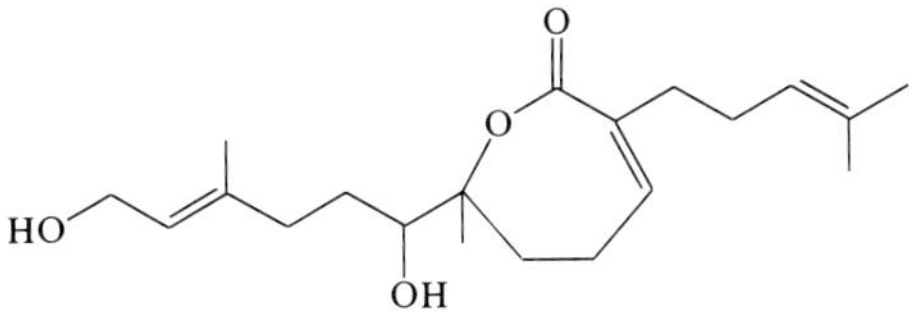

(58)

An unstable compound, toxin 1 (**59**), has been isolated from the fungus *Alternaria citri* (*8*). Several derivatives of (**59**), including the acetate and phenylboronate, were prepared in an attempt to stabilise this compound (*78*). The decarboxylated product (**60**) is obtained on acidic or basic hydrolysis of (**59**). An X-ray structure determination of (**60**) was used in conjunction with ^{1}H-NMR, ^{13}C-NMR and CD spectral data to assign the absolute stereochemistry of (**60**) and hence that of (**59**). Four minor toxins with similar structures to toxin 1 but possessing an α-pyrone nucleus have also been isolated from *A. citri* (*79*).

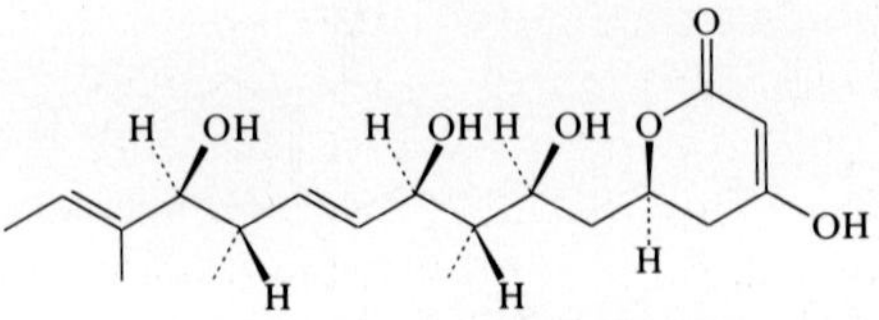

(59)

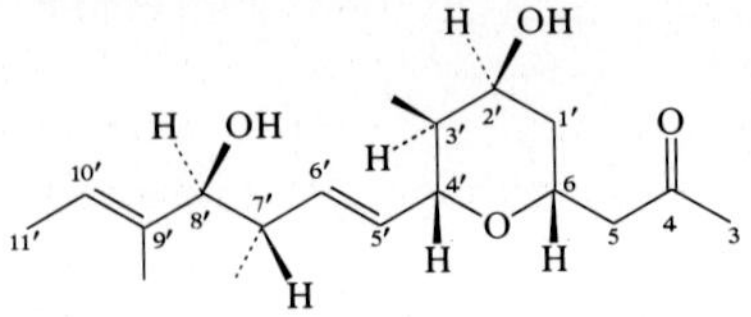

(60)

3.3. 6-Aryl-5,6-dihydro-α-pyrones

The simplest 6-aryl-5,6-dihydro-α-pyrone is psilotin (**61**), first isolated from *Psilotum nudum* (*80*) and then from *Tmesipteris tannensis* (*81*), both members of the family Psilotaceae. Hydrolysis of (**61**) yielded the aglycone psilotinin (**62**) and D-glucose. Chromium trioxide oxidation of (**62**) gave p-hydroxybenzoic acid which indicated that a 1,4-disubstituted benzene ring was present. Synthesis of racemic (**62**) confirmed the structure. Although (**62**) has a single asymmetric centre at C-6, psilotinin is surprisingly optically inactive. Racemisation is thought to occur during the enzymic hydrolysis of (**61**). 3″-Hydroxypsilotin (**63**), a minor compound from *P. nudum*, has been shown to undergo similar racemisation (*82*). The (6*S*)-configuration was established for psilotin from CD studies and by total synthesis of (**61**) and its epimer (*83*). No CD data are reported for (**63**) but from the stereochemistry of (**61**) the (6*S*)-configuration can be presumed.

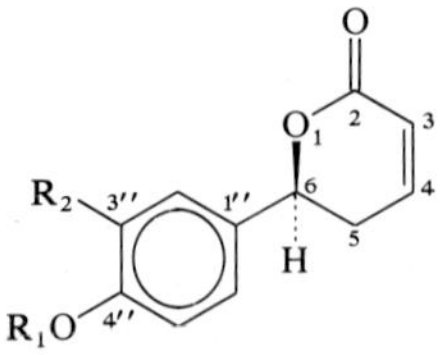

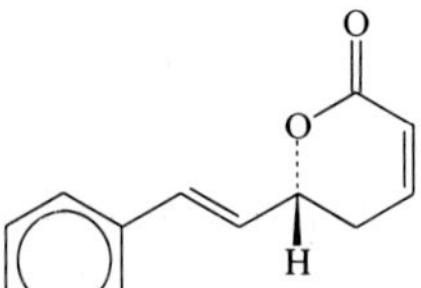

(**61**) $R_1 = \beta$-D-glucosyl, $R_2 = H$
(**62**) $R_1 = R_2 = H$
(**63**) $R_1 = H$, $R_2 = OH$

(**64**)

Goniothalamin (**64**), a 5,6-dihydro-α-pyrone with a 6-styryl substituent found in four species of plants from two different families, was

References, pp. 29–35

first extracted by HLUBUCEK and ROBERTSON (*84*) from the bark of *Cryptocarya caloneura*. ^{1}H-NMR and other physical methods revealed the presence of an αβ-unsaturated-δ-lactone ring and a monosubstituted benzene ring which was confirmed by oxidation to benzoic acid. The *trans* configuration of the exocyclic double bond was assigned from the IR absorbance at 973 cm^{-1}.

The absolute stereochemistry at C-6 in (**64**) has been uncertain. The (6*S*)-stereochemistry was proposed by HLUBUCEK and ROBERTSON from the oxidative ozonolysis of goniothalamin to L-malic acid, isolated as its crystalline xanthate. Their assignment survived for twelve years despite its lone contradiction of Snatzke's rules (*85, 86*). This anomalous situation prompted MEYER (*87*) to synthesise the naturally occurring (+)-(*R*)- (**64**) and its (−)-(*S*)-enantiomer from (−)-(*E,R*)- and (+)-(*E,S*)-3-hydroxy-5-phenyl-4-pentenoic acid respectively.

Goniothalamus species (Annonaceae) are a rich source of 6-aryl-5,6-dihydro-α-pyrones. Goniothalamin has been isolated from three species, *G. andersonii, G. macrophyllus* and *G. malayanus* (*88*). *G. sesquipedalis* and *G. grifithii* have yielded four analogues of (**64**) differing in the hydroxylation and/or acetylation patterns at C-1′ and C-2′, namely goniodiol (**65**), goniodiol monoacetate (**66**), goniodiol diacetate (**67**) and goniotriol (**68**) (*89*). The structures of these compounds are based on ^{1}H-NMR and MS measurements. Goniothalamin (**64**) can be assumed to be the logical precursor of these compounds but the (6*S*)-stereochemistry assigned to them on the assumption that they are derived from goniothalamin is incorrect in the light of MEYER's work.

(**65**) R$_1$ = R$_2$ = R$_3$ = H
(**66**) R$_1$ = R$_3$ = H, R$_2$ = Ac
(**67**) R$_1$ = R$_2$ = Ac, R$_3$ = H
(**68**) R$_1$ = R$_2$ = H, R$_3$ = OH

(**69**)

The stereochemistry at C-1′ and C-2′ was shown by interconversion to be identical in compounds (**65–68**). From the size of the coupling constants of the H-1′ and H-2′ protons (7.5–8.2 Hz) and consideration of *gauche* interactions between the phenyl and lactone rings the configuration (**65**) was proposed for goniodiol. The (1′*S*)- and (2′*S*)-configuration in all four of these compounds was advanced from a biosynthetic hypothesis involving epoxidation of goniothalamin and subsequent enzymatic opening of the *trans* epoxide to the diol (**65**). The epoxide

(69), which has recently been isolated from *G. macrophyllus* (90), is obtained together with its diastereomer by epoxidation of (64). The relative stereochemistry of these epoxides could be deduced from the size of their H-1', H-2' and H-1', H-6 coupling constants, but the absolute stereochemistry of the natural epoxide (69) should be corrected to (6*R*,1'*S*,2'*S*) because of the incorrect (6*S*)-assignment. Conversion of (69) to (65) would settle any inconsistencies in the absolute stereochemistry of compounds (65–68).

The genus *Cryptocarya* has also yielded a novel 6-aryl-5,6-dihydro-α-pyrone with an exocyclic double bond at the 3'-position. Cryptocaryalactone was first isolated by GOVINDACHARI and PARTHASARATHY (91) from *C. bourdilloni* Gamb. and subsequently from *C. moschata* (92). Structure (70) was proposed from ^{1}H-NMR and MS data. The stereochemistry at C-6 and C-2' was not defined and the *trans* configuration of the C-3' double bond was once again assumed from an IR absorbance at 965 cm^{-1}. Naturally occurring (+)-(70) and its epimer have been synthesised from the protected asymmetric aldehyde (71) (93). CD measurements on the two diastereomers, (+)-(6*R*,2'*S*) and (−)-(6*S*,2'*S*) confirmed the (6*R*)-stereochemistry of (+)-(70).

(70)   (71)

A large number of 6-aryl-5,6-dihydro-α-pyrones, methoxylated at C-4, have been isolated from the tropical shrub *Piper methysticum* Forst. (Piperaceae), widely known in the South Pacific as kava, kawa or yanqona. It is an important folk medicine and forms the basis of a ceremonial and social drink much favoured by the Polynesians.

The chemistry of the kawa lactones or piperolides (72–79) has been extensively described in the literature. The structure elucidation of kawain (72), methysticin (73), dihydrokawain (74) and dihydromethysticin (75) has been thoroughly reviewed (1). The identical stereochemistry at C-6 in all four of these compounds follows from CD measurements (94). Hydrogenation of the exocyclic double bond results in a stereochemical order reversal about C-6. Accordingly (72) and (73) possess the (6*R*)-configuration and (74) and (75) the (6*S*)-configuration. The equatorial orientation of the C-6 substituent has been confirmed by ^{1}H-NMR studies of compounds (72–79) and this gives further credibility to the results from the CD determinations (95). Support for the stereochemical assignment of methysticin, (+)-(**73**), has been provided by chemical degradation to D-malic acid which was isolated as its *bis*-phenylhydrazide (*96*).

(**72**) R₁=R₂=H
(**73**) R₁+R₂=—O—CH₂—O—

(**74**) R₁=R₂=R₃=H
(**75**) R₁=R₂=—O—CH₂—O—, R₃=H
(**76**) R₁=R₂=H, R₃=OH
(**77**) R₁=OMe, R₂=R₃=H
(**78**) R₁=OH, R₂=OMe, R₃=H
(**79**) R₁=R₂=OMe, R₃=H

In addition to ¹H-NMR and MS evidence for the structure of dihydrokawain-5-ol (**76**) (*97*), the positive Cotton effect suggested the (*S*)-configuration for both chiral centres which was confirmed by the synthesis of naturally occurring (+)-(**76**) from (**72**) (*98*). Oxidation of (**72**) with selenium dioxide yielded the two diastereomers of kawain-5-ol which were separated by column chromatography. Surprisingly, catalytic hydrogenation of the (6*R*,5*S*)-diastereomer gave only one product, (+)-dihydrokawain-5-ol (**76**) with almost total retention of optical purity.

The development of a GC-MS method for the separation and identification of the kawa lactones from *P. methysticum* Forst. gave a new lactone, (+)-5,6,1′2′-tetrahydroyangoin (**77**), which was later also isolated by column chromatography (*99*).

Finally, column chromatography of a methanolic extract of the roots of *P. methysticum* Forst. afforded two compounds, 4″-hydroxy-3″-methoxydihydrokawain (**78**) and 4″,3″-dimethoxydihydrokawain (**79**), whose structures and absolute stereochemistry were determined by ¹H-NMR, MS and CD methods (*100*). Compounds (**74**), (**75**), (**77**) and (**79**) have also been isolated from a Brazilian species *Aniba gigantigolia* (Lauraceae) (*101*).

4. Physical Methods of Structure Determination

¹H- and ¹³C-NMR spectroscopy, mass spectrometry, UV and IR spectroscopy and increasingly circular dichroism (CD) and X-ray crystallography are used routinely in the structural determination of new

6-substituted 5,6-dihydro-α-pyrones. Detailed spectral data for individual compounds may be found in the papers cited and accordingly this discussion will focus on the more general applications of these spectroscopic techniques. Obvious trends relating to the spectral behaviour of the αβ-unsaturated-δ-lactone ring and variations induced by substitution into this ring are also considered.

4.1. ^{1}H- and ^{13}C-NMR Spectroscopy

^{1}H-NMR spectrometry has played a major role in the structure determination of more than two-thirds of naturally occurring 6-substituted 5,6-dihydro-α-pyrones. Spin decoupling and double resonance experiments have been used extensively to determine chemical shifts and coupling constants but lanthanide shift reagents have been employed only with argentilactone (**35**) (*61*) to resolve complex overlapping ^{1}H-resonances. High-resolution multiple pulse techniques have been applied in the interpretation of the spectra of boronolide (*32*), hyptolide (*66*) and synrotolide (*72*).

(**80**)

Additive shielding parameters (*102*) are routinely used to predict the chemical shift of olefinic protons in the ^{1}H-NMR spectra of unsaturated compounds. Proton H-3 in structure (**80**) resonates at δ 5.9–6.1 and is coupled to H-4 ($J = 9.7$–10 Hz), indicative of a *cis* olefinic function adjacent to a carbonyl group. Proton H-3 is also coupled by long-range coupling (*33, 61, 66*) to the two protons attached to C-5 ($J_{3,5ax}$ = ca. 1 Hz, $J_{3,5eq} = 2$–3 Hz). Methoxylation at C-4 results in shielding of H-3 which then resonates upfield at δ 5.13–5.18 (*95*). The deshielding of H-4 (δ 6.78–7.05) relative to H-3 is typical of a proton attached to the β-carbon of an αβ-unsaturated carbonyl chromophore. The signal for H-4 appears as a triplet of doublets from its coupling to H-3 and also to H-5ax ($J_{4,5ax} = 2$–4 Hz) and H-5eq ($J_{4,5eq} = 4$–6 Hz). These *J*-values have been used to determine the relative stereochemistry of substituents at C-5 (*44, 45, 49, 54*).

The two allylic protons at C-5 are not equivalent and exhibit typical geminal coupling ($J_{5ax,5eq} = 15$–19 Hz) and a coupling to H-4 and to

H-6 ($J_{5ax,6} = 9$–12 Hz; $J_{5eq,6} = 3$–6 Hz). These two protons resonate as a complex multiplet with chemical shifts between δ 2.3 and δ 2.8. Methoxylation at C-4 has a minimal deshielding effect on the chemical shifts of H-5ax and H-5eq (*95*). Hydroxylation or acetylation at C-5 causes a downfield shift of the remaining axial or equatorial H-5 proton to δ 4.2–5.4 (*44, 45, 49*).

A study of parasorbic acid (**5**) suggests that the two H-5 protons make dihedral angles of about 160° and 40° with H-6 ($J_{5ax,6} = 10.3$ Hz and $J_{5eq,6} = 5.4$ Hz) and hence the C-6 methyl group is *pseudo-* equatorially oriented in the most stable conformation (*103*). The H-5,H-6 coupling constants have also been used to assign a similar orientation to the C-6 side chains of compounds (**22**) (*44*), (**23**) (*45*), (**24**) (*49*), (**35**) (*61*), (**39**) (*66*) and (**72–76**) (*95*). The H-6 chemical shift (δ 4.2–5.1) is determined by the nature of the substituent at C-6. Oxygenation of the side chain results in a number of contiguous chiral centres whose relative stereochemistry has been determined by application of the nuclear Overhauser effect (**78**).

Important structural information is also provided by ^{13}C-NMR spectrometry. The chemical shift of the C-2 carbonyl carbon atom is usually at δ 160–169. Carbon atom C-3 resonates at δ 120–122 but methoxylation at C-4 causes this signal to shift to δ 89–94. The signal at C-4 (δ 144–148) is conversely shifted *ca.* 30 ppm downfield by the presence of this methoxy group. The C-5 carbon atom resonates at δ 25–30 and this signal is only slightly affected by methoxylation at C-4. Oxygenation at C-5 causes a downfield shift of the ^{13}C-signal by about 40 ppm and has little effect on the C-4 chemical shift but the C-6 resonance, normally at δ 72–78, is shifted downfield by about 35 ppm (*104*).

4.2. Mass Spectrometry

Mass spectrometry unlike NMR spectrometry has been used predominantly as a supplementary technique for determining the chemical structure of 6-substituted 5,6-dihydro-α-pyrones. The αβ-unsaturated-δ-lactone ring undergoes a series of fragmentations to yield ions of substantial diagnostic value. The fragmentation modes of eight 6-alkyl-5,6-dihydro-α-pyrones have been studied by URBACH, STARK and NOBUHARA (*105*). The major mass spectral fragments of the simplest 6-substituted 5,6-dihydro-α-pyrone, parasorbic acid (**5**), are shown in *Scheme 1* (*105, 106*). The two ions at *m/z* 97 and *m/z* 68 correspond to the facile α-cleavage of the C-6 side chain which is characteristic of this group of compounds. The ion at *m/z* 68 also formed from further fragmentation of the *m/z* 97 ion (*27, 61, 91*). The structure of the former

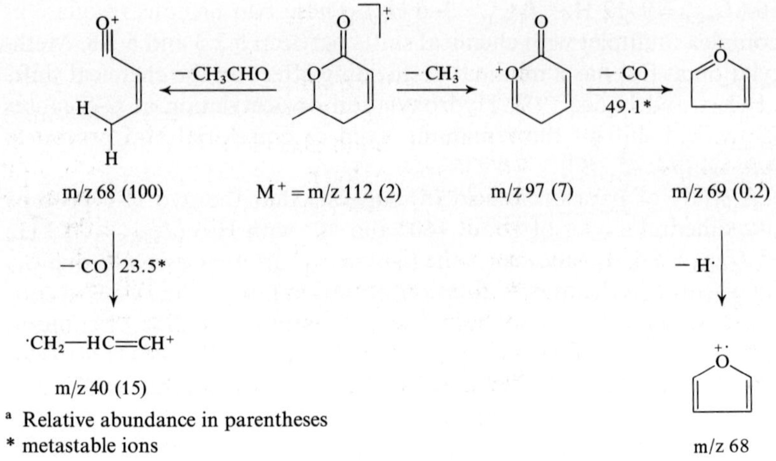

Scheme 1. Major mass spectral fragments of parasorbic acid [a]

fragment has been a source of well-documented controversy (*107*) and both proposed structural representations are given in *Scheme 1*. The cleavage of the C-6 substituent also affords relatively abundant ions in the mass spectrum of multi-substituted 5,6-dihydro-α-pyrones as represented by osmundalactone (**22**) (*Scheme 2*) (*44*).

Finally, 6-substituted 5,6-dihydro-α-pyrones, with the exception of some polyhydroxylated compounds (*33*, *44*, *78*), exhibit a molecular ion in their mass spectra. These compounds undergo thermal decomposition but this problem can be alleviated by trimethylsilylation.

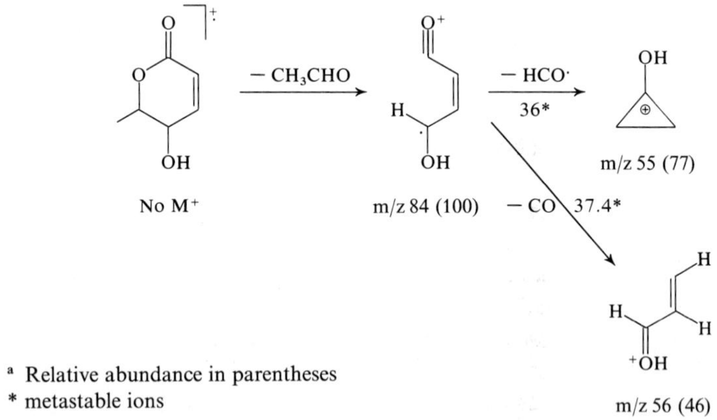

[a] Relative abundance in parentheses
* metastable ions

Scheme 2. Major mass spectral fragments of osmundalactone[a]

References, pp. 29–35

4.3. Infrared and Ultraviolet Spectroscopy

The IR spectra of 6-substituted 5,6-dihydro-α-pyrones exhibit a strong absorption band at 1710–1730 cm^{-1} due to the αβ-unsaturated carbonyl group. The lactone double bond absorbs at 1590–1640 cm^{-1}, and although this absorption is usually much weaker than that of the carbonyl group, its intensity is enhanced by oxygenation at either C-4 or C-5 (*37, 44*). A further absorption at 1555 cm^{-1} displayed by lachnelluloic acid (**27**) (*57*) and related compounds (**28–30**) (*58*) has been interpreted as arising from the strongly chelated, enolized β-diketone (**81**) (*57*). A tenuous *trans* configuration for the exocyclic double bond in compounds (**37**) (*33*), (**38**) (*33*), (**41**) (*65*), (**64**) (*84*) and (**70**) (*91*) has been assigned from the infrared absorption at *ca.* 965 cm^{-1}.

(**81**)

The UV absorption maxima (λ) of monosubstituted 6-alkyl- and 6-alkenyl-5,6-dihydro-α-pyrones are normally between 200–215 nm (ε = 800–12000). Methoxylation at C-4 induces a bathochromic shift of approximately 20 nm (*37*). Hydroxylation or acetylation at C-5 (*44*) has no effect on the value of λ but acylation at C-3 generates a further absorption at 274 nm (ε = 1100) (*57*). 6-Aryl-5,6-dihydro-α-pyrones, e.g. psilotin (**61**), also absorb at 275–285 nm. The extended chromophore of the styryl substituent in compounds (**64**) (*84*), (**70**) (*91*), (**72**) and (**73**) (*1*) is responsible for three characteristic strong UV absorptions at 250 nm (ε = *ca.* 19000), 283 nm (ε = *ca.* 16000) and 292 nm (ε = *ca.* 12000).

4.4. Circular Dichroism

CD and ORD data of 6-substituted 5,6-dihydro-α-pyrones are presented in *Table 1*.

In the CD or ORD spectra of αβ-unsaturated-δ-lactones the carbonyl group exhibits a distinct n→π* Cotton effect near 260 nm (*85, 86, 108*). The sign of the CD curve is considered by SNATZKE (*85*) to denote either conformation (**82**) or (**83**) for the lactone ring. This interpretation, widely known as Snatzke's rules, is based on the CD spectrum of parasorbic acid (**5**) and the assumption that the co-planar-

Table 1. *The n→π* CD and ORD Data for 6-substituted-5,6-dihydro-α-pyrones*

Compound	No.	CD Δε	n→π* λ nm	ORD Φ	n→π* λ nm	Stereochemistry at C-6 implied from Snatzke's rules
Anamarine	42	+[a]	260	–	–	(R)
Argentilacetone	35	+2	256	–	–	(R)
Asperline	23	+2	265	–	–	(R)
Aspyrone	25	−4.8	263	−1341	263	(R)
Boronolide	11	+2.48	256	–	–	(R)
C_{26} Lactone	14	−1.32	252	–	–	(S)
C_{26} Lactone	15	−0.59	253	–	–	(S)
Cryptocaryalactone	70	+1.17	265	–	–	(R)
Dihydrokawain	74	+11.2	246	–	–	(S)
Dihydrokawain-5-ol	76	+11.5	247	–	–	(S)
Dihydromethysticin	75	+11.0	246	–	–	(S)
3″,4″-Dimethoxy-dihydrokawain	79	+13.13	247	–	–	(S)
Goniothalamin	64	+6.5	253	–	–	(R)
4″-Hydroxy-3″-methoxydihydrokawain	78	+8.58	247	–	–	(S)
Hyptolide	39	+2.3	257	–	–	(R)
Kawain	72	+8.3	249	–	–	(R)
Massoilactone	8	–	–	−8900	265	(R)
Methysticin	73	+8.9	242	–	–	(R)
Olguine	43	+[a]	270	–	–	(R)
Osmundalactone	22	+3.86	263	–	–	(S)
Parasorbic acid	5	+2.25	262	+7400	265	(S)
Pestalotin	16	−7.90	243	–	–	(S)
Phomalactone	24	+0.7	265	–	–	(R)
Psilotin	61	−3.3	270	–	–	(S)
Synrotolide	45	+2.45	266	–	–	(R)
Toxin 1	59	−7.75	246	–	–	(R)

[a] Value not cited in Literature.

ity through the lactone ester group of saturated δ-lactones is retained in αβ-unsaturated-δ-lactones.

The assignment of conformation (**82**) (R = Me) to parasorbic acid by Snatzke arises from the known stereochemistry (*14, 15*) and the *pseudo*-equatorial orientation of the methyl group established from ^{1}H-NMR studies (*103*). Conversely therefore, Snatzke's rules can be used to assign the absolute stereochemistry at C-6 if the sign of the Cotton effect and the orientation of the C-6 substituent are known.

References, pp. 29–35

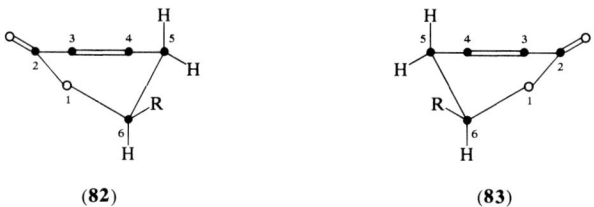

(82) (83)

Although the H-5,H-6 coupling constants remain crucial parameters for the unequivocal confirmation of the orientation of the C-6 substituent, a *pseudo*-equatorial orientation of the C-6 side chain can be inferred from consideration of steric factors. With the exception of the two C_{26}-lactones (14) and (15) (*35*) the absolute stereochemistry at C-6 implied by Snatzke's rules (*Table 1*) has been confirmed by either chemical degradation, synthesis or biosynthetic relationships. Therefore, in the absence of any evidence to the contrary, a *pseudo*-equatorial orientation of the C-6 substituent can be assumed for all 6-substituted 5,6-dihydro-α-pyrones.

The conformation of the αβ-unsaturated-δ-lactone ring has been the subject of considerable speculation (*86, 109, 110*). BEECHAM (*86*) has proposed that coplanarity through the —C—CO—O—C group, as assumed by SNATZKE, is not normal in αβ-unsaturated-δ-lactones and suggests that the lactone ring adopts the minimum energy enantiomeric conformations (84) and (85), in which the C-6 atom is displaced by *ca*. 0,5 Å from a least squares plane containing C-2, C-3, C-4 and C-5. Atom O–1 is slightly displaced towards C-6. The X-ray crystallographic analyses of boronolide (11) (*31*), hyptolide (39) (*66*) and goniothalamin (64) (*111*) support this proposal. However, despite this deviation from the original assumption, the empirical interpretation of Snatzke's rules is still valid. BEECHAM (*86*) has proposed that the positive or negative Cotton effect generated by the allylic oxygen atom can be related respectively to a right handed or left handed chirality of the oxygen-olefinic bond. Evidence in support of this proposal is provided by the fungal metabolites asperline (23), phomalactone (24) and aspyrone (25). The chirality at C-5 in these compounds indicated

(84) (85)

by the positive $\pi \to \pi^*$ CD peak agrees with that obtained by NMR studies in conjunction with the conformation of the lactone ring determined by Snatzke's rules.

5. Biosynthesis

Studies on the biosynthesis of only three compounds, parasorbic acid (**5**), aspyrone (**25**) and psilotin (**61**), have been reported. The first biosynthetic investigation of a 6-substituted-5,6-dihydro-α-pyrone was made by CROMBIE and FIRTH (*112*) on parasorbic acid in the rowan berry (*Sorbus aucuparia* L.). The structure of (**5**) suggests that L-rhamnose (**86**) is a possible 6-deoxy-sugar precursor in which the (6S)-configuration of (**5**) arises directly from the (5S)-configuration of (**86**). However, L-rhamnose is absent from rowan berries and the major monosaccharides are D-glucose (**87**) and D-sorbitol. Although (**86**) has a (5R)-configuration it may still be a precursor of (**5**). This hypothesis was tested by injection of [1-^{14}C], [2-^{14}C]- and [6-^{14}C]-D-glucose into ripening rowan berries. The biosynthesis of (**5**) via an acetate-malonate pathway was also investigated by means of sodium [1-^{14}C]- and [2-^{14}C]-acetate and sodium [1-^{14}C]-malonate. The substantially higher isotope incorporations as well as the distributions of the ^{14}C-label in the C-1, C-5 and C-6 fragments of (**5**) in the latter experiment, suggests that the acetate-malonate pathway is the major biosynthetic route.

(**86**)   (**87**)

The extensive investigations on the biosynthesis of aspyrone (**25**) by the fungus *Aspergillus melleus* have been summarized by BRERETON, GARSON and STAUNTON (*113*). The polyketide origin of aspyrone was established by means of ^{14}C- and ^{13}C-labelled precursors. Further experiments with doubly labelled ^{13}C-acetate showed that (**25**) is formed from three intact acetate units and three carbons derived from cleaved acetate units. The C-3 and C-7 ^{13}C-coupling constant (61 Hz) shows that the bond between these two atoms is formed by head-to-head linkage of two acetate units. In addition, the long range ^{13}C-coupling constant (6.2 Hz) between C-2 and C-7 indicative of two carbon atoms derived from the same acetate unit. It was therefore suggested that

References, pp. 29–35

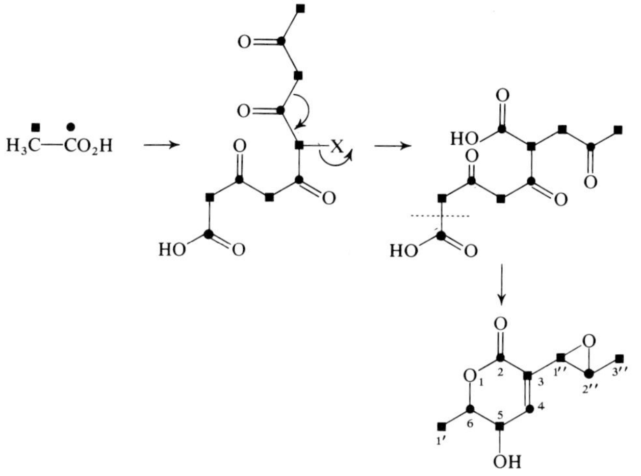

Scheme 3. The incorporation of ^{13}C labelled acetate in the biosynthesis of aspyrone

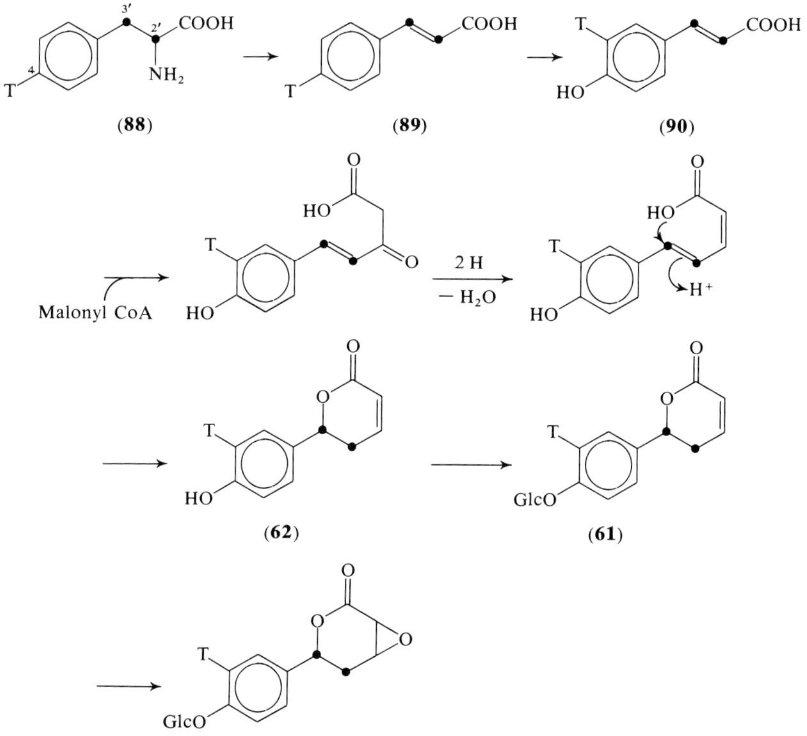

Scheme 4. The proposed biosynthesis of psilotin from phenylalanine

the biosynthesis of (**25**) proceeds *via* a Favorskii-type rearrangement of the linear polyketone chain (*Scheme 3*).

Phenylalanine (**88**) is the major precursor in the biosynthesis of psilotin (**61**) (*114*). The biosynthetic pathway (*Scheme 4*) was proposed from isotope incorporation studies with (*RS*)-[2′,3′-$^{13}C_2$]- and (*S*)-[1′-^{14}C]-phenylalanine. The first step is the deamination of (**88**) to cinnamic acid (**89**). Tritium-labeled phenylalanine was used to monitor the expected NIH shift which occurs in the conversion of (**89**) to *p*-coumaric acid (**90**). Chain extension of (**90**) occurs by the addition of a single acetate unit as malonyl CoA. The proposed pathway is supported by the ^{13}C-NMR spectra of the isolated (**61**) and psilotin epoxide (**91**) which show that the ^{13}C-isotopes from [2′,3′-$^{13}C_2$]-phenylalanine are retained at C-5 and C-6. The structural similarity between (**62**) and the kawa lactones suggests a common biosynthetic pathway and kawain may be biosynthesised from (**90**) by the addition of two acetate units.

6. Biological Activity

A number of 6-substituted 5,6-dihydro-α-pyrones are allelochemicals. Parasorbic acid (**5**) (*115*), psilotin (**61**) (*3*) and cryptocaryalactone (**70**) (*92*) inhibit seed germination and plant growth. The inhibitory effects of (**61**) are reversed by glutathione, other thio-compounds and gibberellin A_3, suggesting that (**61**) forms part of a growth regulatory system (*3*).

6-Substituted 5,6-dihydro-α-pyrones isolated from fungi exhibit a variety of biological activities. Compounds (**23**) and (**24**) are active against bacteria (*46, 50*). Pestalotin (**16**) enhances the growth-stimulative effect of gibberellic acid in plants (*37*), while conversely toxin 1 (**59**) (*8*) and alternaric acid (**26**) (*5*) are phytopathogenic. Compound (**59**) is responsible for brown spot disease on lemons and Rangpur limes while (**26**) has been shown to cause the collapse of tissues and wilting in plants from the families Solanaceae, Cruciferae and Compositae. Alternaric acid and the other fungal metabolites acylated at C-3 are remarkably specific antifungal toxins. Alternaric acid inhibits germination of the spores of *Absidia glauca*, *Myrothecium verrucia* and *Stachybotrys atra* at very low concentrations (*5*), lachnelluloic acid (**27**) is antagonistic towards Dutch elm disease and wood rotting fungi (*57*), while the podoblastins A, B, C (**28–30**), have been reported to possess specific antifungal activity against rice blast (*58*). The higher plant metabolite goniothalamin (**64**) also exhibits antifungal activity (*88*).

Osmundalactone (**22**) is the only naturally occurring 6-substituted 5,6-dihydro-α-pyrone known to possess insect antifeedant properties; it inhibits the feeding of the larvae of the yellow butterfly *Eurema hecabe mandarina* on the fern *Osmunda japonica* Thunberg (*4*). Interestingly, two species of ants from the genus *Componotus* employ massoilactone (**8**) as a defensive substance (*27*). The kawa lactones (**72–77**) which are the active principles from the root, rhizome and lower stem of *Piper methysticum* Forst possess a range of pharmacological activities (*116*). These include smooth muscle relaxant, local anaesthetic, protection against electro- and chemo-shock, anti-inflammatory, antipyretic, antimycotic and antioedemic activities.

Finally, compounds closely related to the kawa lactones also have interesting physiological properties, e.g. goniothalamin from *Goniothalamus macrophyllus* exhibits CNS activity (*88*). *G. macrophyllus* is used as an abortifacient in rural areas of Northern Malaysia and both (**64**) and the epoxide (**69**) are reported to induce foetal abnormalities in mice (*90*).

References

1. Mors, W.B., M.T. Magalhães, and O.R. Gottlieb: Naturally Occurring Aromatic Derivatives of Monocyclic α-Pyrones. Fortschr. Chem. organ. Naturstoffe **20**, 131 (1962).
2. Adityachaudhury, N., and A.K. Das: Recent Advances in the Chemistry of Naturally Occurring 2-Pyrone derivatives. J. Sci. Indust. Res. (India) **38**, 265 (1979).
3. Siegel, S.M.: Inhibitory Activity of the Phenolic Glucoside Psilotin and its Reversal by Gibberellic Acid and Thiols. Phytochem. **15**, 566 (1976)
4. Numata, A., K. Hokimoto, T. Takemura, T. Katsuno, and K. Yamamoto: Plant Constituents Biologically Active to Insects. V. Antifeedants for the Larvae of the Yellow Butterfly *Eurema hecabe mandarina*, in *Osmunda japonica*. Chem. Pharm. Bull (Japan) **32**, 2815 (1984).
5. Brian, P.W., P.J. Curtis, H.G. Hemming, C.H. Unwin, and J.M. Wright: Alternaric Acid, a Biologically Active Metabolic Product of the Fungus *Alternaria solani*. Nature **164**, 534 (1949).
6. Stampwala, S.S., R.H. Bunge, T.R. Hurley, N.E. Willmer, A.J. Brankiewicz, C.E. Steinman, T.A. Smitka, and J.C. French: Novel Antitumour agents CI-920, PD 113270 and PD 113271. II. Isolation and Characterization. J. Antibiot. **36**, 1601 (1983).
7. McGahren, W.J., G.A. Ellestad, G.O. Morton, M.P. Kunstmann, and P. Mullen: New Fungal Lactone, LL-P880β and a new Pyrone LL-P880γ from a *Penicillium* sp. J. Organ. Chem. (USA) **38**, 3542 (1973).
8. Gardener, J.M., Y. Kono, J.H. Tatum, Y. Suzuki, and S. Takeuchi: Plant Phototoxins from *Alternaria citri*. The Major Toxin Specific for Rough Lemon Plants. Phytochem. **24**, 2861 (1985).
9. Chmielewska, I., J. Cieslak, K. Gorczynska, B. Kontnik, and K. Pitakowska: Structure of Yangonine. Ultraviolet and Infrared Spectrographic Studies. Tetrahedron **4**, 36 (1958).

10. RIGAUDY, J., and S.P. KLESNEY: IUPAC Nomenclature of Organic Chemistry, Sections A-H, pp. 171 and 203. Pergamon Press, 1979.
11. HOFMANN, A.W.: Liebigs Ann. Chem. **110**, 129 (1859).
12. DOEBNER, O.: Ber. dtsch. chem. Ges. **27**, 344 (1894).
13. KUHN, R., and D. JERCHEL: 2-Hexen-4, 1-olide and 2-Hexen-5, 1-olide. Constitution of the Parasorbic Acid from the Volatile Oil of Ripe Rowanberries. Ber. dtsch. chem. Ges. **76B**, 413 (1943).
14. KUHN, R., and K. KUM: The Absolute Configuration of Sorbin Oil. Chem. Ber. **95**, 2009 (1962).
15. LUKĚS, R., J. JARÝ, and J. NEMEC: Lactones. VII. 4,6-Dideoxy-L-ribo-hexano-1,5-lactone and the Absolute Configuration of Parasorbic Acid. Collect. Czech. Chem. Comm. **27**, 735 (1962).
16. LICHTENTHALER, F.W., F.D. KLINGLER, and P. JARGLIS: Simple Synthesis of (S)-Parasorbic Acid and other (5S)-Hydroxy Sixcarbon Synthons from L-Rhamnose. Carbohydr. Res. **132**, C1 (1984).
17. TSCHESCHE, R., H.J. HOPPE, G. SNATZKE, G. WULFF, and H.W. FEHLHABER: Glycosides with Lactone – Forming Aglycons. III. Parasorbiside, the Glycosidic Precursor of Parasorbic Acid, from Mountain Ash Berries. Chem. Ber. **104**, 1420 (1971).
18. OHLOFF, G.: Recent Developments in the Field of Naturally Occurring Aroma Components. Fortschr. Chem. organ. Naturstoffe **35**, 431 (1978).
19. FUJIMORI, T., R. KASUGA, H. MATSUSHITA, H. KANEKO, and M. NOGUCHI: The Aroma of Burley Tobacco. Part I. Neutral Aroma Constituents in Burley Tobacco. Agric. Biol. Chem. **40**, 303 (1976).
20. MEIJR, T.M.: Essential Oil of Massoy Bark. Rec. trav. chim. Pays-Bas **59**, 191 (1940).
21. ABE, S., and K. SATO: The Essential Oil of Massoia. III. Structure of Massoia Lactone. J. Chem. Soc. Japan **75**, 952 (1954).
22. MORI, K.: Absolute Configuration of (−)-Massoilactone as Confirmed by a Synthesis of its (S)-(+)-Isomer. Agric. Biol. Chem. **40**, 1617 (1976).
23. PIRKLE, W.H., and P.E. ADAMS: Enantiomerically Pure Lactones. 3. Synthesis of and Stereospecific Conjugate Additions to $\alpha\beta$-Unsaturated Lactones. J. Organ. Chem. (USA) **45**, 4117 (1980).
24. KAISER, R., and D. LAMPARSKY: Das Lacton der 5-Hydroxy-cis-2, Cis-7-decadiensäure und weitere Lactone aus dem Absud der Blüten von *Polianthes tuberosa* L. Tetrahedron Letters 1659 (1976).
25. BAYER, E.: Quality and Flavour by Gas Chromatography. J. Gas Chromatogr. **4**, 67 (1966).
26. HASHIZUME, T., N. KIKUCHI, Y. SASAKI, and I. SAKATA: Constituents of Cane Molasses. III. Isolation and Identification of (−)-2-deceno-5-lactone (Massoilactone). Agric. Biol. Chem. **32**, 1306 (1968).
27. CAVILL, G.W.K., D.V. CLARK, and F.B. WHITFIELD: Insect Venoms, Attractants, and Repellents. XI. Massoilactone from two Species of Formicine Ants, and some Observations on Constituents of the Bark Oil of *Cryptocarya massoia*. Austral. J. Chem. **21**, 2819 (1968).
28. BOHLMANN, F., and A. SUWITA: Ein neues Bisabolen-Derivat und ein neues Dihydro-Kaffeesäure-Derivat aus *Tarchonanthus trilobus*. Phytochem. **18**, 677 (1979).
29. NAKATA, T., N. HATA, K. IIDA, and T. OISHI: Determination of Stereostructure of Naturally Occurring α,β-Unsaturated δ-Lactone Derivates Through a Stereoselective Synthesis. Tetrahedron Letters **28**, 5661 (1987).
30. FRANCA, N.C., and J. POLONSKY: Sur la Structure du Boronolide, Isolé du *Tetradenia fruiticosa* Benth. C.R. hebd. seances Acad. Sci. **273** C, 439 (1971).
31. KJAER, A., R. NORRESTAM, and J. POLONSKY: Boronolide. Structure and Stereochemistry (X-ray Analysis). Acta Chem. Scand. **B39**, 745 (1985)

32. DAVIES-COLEMAN, M.T., and D.E.A. RIVETT: Stereochemical Studies on Boronolide, an α-Pyrone from *Tetradenia barberae*. Phytochem. **26**, 3047 (1987).
33. VAN PUYVELDE, L., S. DUBÉ, E. UWIMANA, C. UWERA, R.A. DOMISSE, E.L. ESMANS, O. VAN SCHOOR, and A.J. VLIETINCK: New α-Pyrones from *Iboza riparia*. Phytochem. **18**, 1215 (1979).
34. VAN PUYVELDE, L., N. DE KIMPE, S. DUBÉ, M. CHAGNON-DUBÉ, Y. BOILY, F. BORREMANS, N. SCHAMP, and M.J.O. ANTEUNIS: 1,2-Dideacetylboronolide, an α-Pyrone from *Iboza riparia*. Phytochem. **20**, 2753 (1981).
35. HERZ, W., and G. RAMAKRISHNAN: δ-Lactones of Polyhydroxy-C_{26} acids in *Eupatorium pilosum*. Phytochem. **17**, 1327 (1978).
36. Private communication from Professor W. Herz.
37. KIMURA, Y., K. KATAGIRI, and S. TAMURO: Structure of Pestalotin, a New Metabolite from *Pestalotia cryptomeriaecola*. Tetrahedron Letters 3137 (1971).
38. ELLESTAD, G.A., W.J. MCGAHREN, and M.P. KUNTSMANN: Structure of a New Fungal Lactone, LL-P880α, from an Unidentified *Penicillium* sp. J. Organ. Chem. (USA) **37**, 2045 (1972).
39. SEEBACH, D., and H. MEYER: Synthesis of (±)-Pestalotin and of Optically Pure (−)-Pestalotin by Asymmetric Synthesis. Angew. Chem. Internat. Edit. **13**, 77 (1974).
40. KIRIHATA, M., K. OHTA, S. YAMAMOTO, I. ICHIMOTO, and H. UEDA: Abstract of Papers Annual Meeting of the Agric. Chem. Sociecty of Japan 1980, p. 211, Fukoka.
41. MASAKI, Y., K. NAGATA, Y. SERIZAWA, and K. KAJI: Facile and Rapid Entry to Functionalized and Optically Active Pyrans from Tartaric Acid by way of 6,8-Dioxabicyclo [3.2.1]octanes. Application to the Synthesis of (−)-(6S,1'S)-Pestalotin. Tetrahedron Letters **25**, 95 (1984).
42. MORI, K., T. OTSUKA, and M. ODA: Synthesis of all the Four Possible Stereoisomers of Pestalotin, a Gibberellin Synergist Isolated from *Pestalotia cryptomeriaecolia* Swada. Tetrahedron **40**, 2929 (1984).
43. STRUNZ, G.M., C.J. HEISSNER, M. KAKUSHIMA, and M.A. STILLWELL: Metabolites of an Unidentified Fungus. A New 5,6-Dihydro-2-pyrone Related to Pestalotin. Canad. J. Chem. **52**, 825 (1974).
44. HOLLENBEAK, K.H., and M.E. KUEHNE: Isolation and Structure Determination of the Fern Glycoside Osmundalin and the Synthesis of its Aglycon Osmundalactone. Tetrahedron **30**, 2307 (1974).
45. ARGOUDELIS, A.D., and J.F. ZIESERL: The Structure of U-13, 933, a New Antibiotic. Tetrahedron Letters 1969 (1966).
46. MIZUBA, S., K. LEE, and J. JIU: Three Antimicrobial Metabolites from *Aspergillus caespitosus*. Canad. J. Microbiol. **21**, 1781 (1975).
47. YAMAMOTO, I., H. SUIDE, T. HEMMI, and T. YAMANO: Antimicrobial αβ-Unsaturated-δ-lactones from Fungi. Takeda Kenkyusho Ho **29**, 1 (1970).
48. JIU, J., S. KRAYCHY, and S.S. MIZUBA: Microbial Production of Antimicrobial 2H-Pyran-2-ones. U.S. Patent 3, 909, 362 (1975).
49. EVANS, R.H., G.A. ELLESTAD, and M.P. KUNSTMANN: Two New Metabolites from an Unidentified *Nigrospora* Species. Tetrahedron Letters 1791 (1969).
50. YAMANO, T., S. HEMMI, I. YAMAMOTO, and K. TSUBAKI: Fermentative Production of the Antibiotic Phomalactone. Japanese Patent 7, 132, 800 (1971).
51. EVANS, R.H., and C.E. HOLMLUND: Fermentative Preparation of Antimicrobial 5,6-Dihydro-5-hydroxy-6-propenyl-2-pyrone and its Derivatives. U.S. Patent 3, 701, 787 (1972).
52. MOORE, J.H., T.P. MURRAY, and M.E. MARKS: Production of 3-(1,2-epoxypropyl)-5,6-dihydro-5-hydroxy-6-methylpyran-2-one by *Aspergillus ochraceus*. J. Agric. Food Chem. **22**, 697 (1974).
53. GARSON, M.J., J. STAUNTON, and P.G. JONES: New Polyketide Metabolites from

Aspergillus melleus. Structural and Stereochemical studies. J. Chem. Soc. Perkin Trans. I 1021 (1984).
54. ROSENBROOK, W., and R.E. CARNEY: New Metabolite from an Unidentified *Aspergillus* species. Tetrahedron Letters 1867 (1970).
55. GROVE, J.F.: Alternaric Acid. Part 1. Purification and Characterisation. J. Chem. Soc. 4056 (1952).
56. BARTELS-KEITH, J.R.: Alternaric Acid. Part III. Structure. J. Chem. Soc. 1662 (1960).
57. AYER, W.A., and J.D.F. VILLAR: Metabolites of *Lachnellula fuscosanguinea* (Rehm). Part 1. The Isolation, Structure Determination, and Synthesis of Lachnelluloic Acid. Canad. J. Chem. **63**, 1161 (1985).
58. MIYAKADO, M., S. INOUE, Y. TANABE, K. WATANABE, N. OHNO, H. YOSHIOKA, and T.J. MABRY: Podoblastin A, B and C. New Antifungal 3-Acyl-4-hydroxy-5,6-dihydro-2-pyrones obtained from *Podophyllum peltatum* L. Chem. Lett. 1539 (1982).
59. PIRKLE, W.H., and J.R. HAUSKE: Broad Spectrum Methods for the Resolution of Optical Isomers. A Discussion of the Reasons Underlying the Chromatographic Separability of Some Diastereomeric Carbamates. J. Organ. Chem. (USA) **42**, 1839 (1977).
60. KORVER, O.: Optical Rotary Dispersion and Circular Dichroism of δ-Lactones. Determination of the Absolute Configuration of (+)5-Decanolide and (+)5-Dodecanolide. Tetrahedron **26**, 2391 (1970).
61. PRIESTAP, H.A., J.D. BONAFEDE, and E.A. RUVEDA: Argentilactone, a Novel 5-Hydroxyacid Lactone from *Aristolochia argentina*. Phytochem. **16**, 1579 (1977).
62. O'CONNOR, B., and G. JUST: Synthesis of Argentilactone and Goniothalamin. Tetrahedron Letters **27**, 5201 (1986).
63. ACHENBACH, H., and J. WITZKE: Studies on Naturally Occurring γ- and δ-Lactones. X. Synthesis of Umuravumbolide and Epiumuravumbolide. Z. Naturforsch., B: Anorg. Chem., Org. Chem. **35**, 1459 (1980).
64. GORTER, K.: Hyptolide a Bitter Principle of *Hyptis pectinata* Poit. Bull. Jard. bot. Buitenzorg 327 (1920).
65. BIRCH, A.J., and D.N. BUTLER: The Structure of Hyptolide. J. Chem. Soc. 4167 (1964).
66. ACHMED, S., T. HOYER, A. KJAER, L. MAKMUR, and R. NORRESTAM: Molecular and Crystal Structure of Hyptolide, a Naturally Occurring αβ-Unsaturated-δ-Lactone. Acta Chem. Scand. **B41**, 599 (1987).
67. ALEMANY, A., C. MARQUEZ, C. PASCUAL, S. VALVERDE, M. MARTINEZ-RIPOLL, J. FAYOS, and A. PERALES: New Compounds from Hyptis. X-ray Crystal and Molecular Structures of Anamarine. Tetrahedron Letters 3583 (1979).
68. ALEMANY, A., C. MARQUES, C. PASCUAL, S. VALVERDE, A. PERALES, J. FAYOS, and M. MARTINEZ-RIPOLL: New Compounds from Hyptis. X-ray Crystal and Molecular Structures of Olguine. Tetrahedron Letters 3579 (1979).
69. DELGADO, G., R. PEREDA-MIRANDA, and A. ROMO DE VIVAR: Structure and Stereochemistry of 4-Deacetoxy-10-epi-olguine, a New δ-Lactone from *Hyptis oblongifolia*. Heterocycles **23**, 1869 (1985).
70. GILLARD, F., and J.J. RIEHL: A Synthetic Approach to (+)-Anamarine. Synthesis of the Precursor of the Side-Chain. Tetrahedron Letters **24**, 587 (1983).
71. VALVERDE, S., A. HERNANDEZ, B. HERRADON, R.M. RABANAL, and M. MARTIN-LOMAS: The Synthesis of (−)-Anamarine. Tetrahedron **43**, 3499 (1987).
72. DAVIES-COLEMAN, M.T., R.B. ENGLISH, and D.E.A. RIVETT: Synrotolide, a New α-Pyrone from *Syncolostemon rotundifolius*. Phytochem. **26**, 1497 (1987).
73. KAZLAUSKAS, R., P.T. MURPHY, R.J. WELLS, and A.J. BLACKMAN: Macrocyclic Enol-Ethers Containing an Acetylenic Group from the Red Alga *Phacelocarpus labillardieri*. Austral. J. Chem. **35**, 113 (1982).
74. HOKANSON, G.C., and J.C. FRENCH: Novel Antitumour Agents CI-920, PD 113270 and PD 113271, 3. Structure Determination. J. Organ. Chem. (USA) **50**, 462 (1985).

75. FUNAISHI, K., K. KAWAMURA, Y. SUGIURA, N. NAKAHORI, E. YOSHIDA, M. OKANISHI, I. UMEZAWA, S. FUNAYAMA, and K. KOMIYAMA: Kazusamycin B, a Novel Antitumour Antibiotic. J. Antibiot. **40**, 778 (1987).
76. BOHLMANN, F., J. JAKUPOVIC, A. SCHUSTER, R.M. KING, and H. ROBINSON: New Melampolides, Kaurene Derivatives and Other Constituents from *Ichthyothere* Species. Phytochem. **21**, 2317 (1982).
77. BOHLMANN, F., J. JAKUPOVIC, A.K. DHAR, R.M. KING, and H. ROBINSON: Two Sesquiterpene and Three Diterpene Lactones from *Acanthospermum australe*. Phytochem. **20**, 1081 (1981).
78. KONO, Y., J.M. GARDNER, K. KOBAYANSHI, Y. SUZUKI, S. TAKEUCHI, and T. SAKURAI: Plant Pathotoxins from *Alternaria citri*. Stereochemistry of the Major and Minor Toxins. Phytochem. **25**, 69 (1986).
79. KONO, Y., J.M. GARDNER, Y. SUZUKI, and S. TAKEUCHI: Plant Pathotoxins from *Alternaria citri*. The Minor ACRL Toxins. Phytochem. **24**, 2869 (1985).
80. MCINNES, A.G., S. YOSHIDA, and G.H.N. TOWERS: A Phenolic Glycoside from *Psilotum nudum* (L) Griseb. Tetrahedron **21**, 2939 (1965).
81. TSE, A., and G.H.N. TOWERS: The Occurence of Psilotin in *Tmesipteris*. Phytochem. **6**, 149 (1967).
82. BALZA, F., A.D. MUIR, and G.H.N. TOWERS: 3'-Hydroxypsilotin, A Minor Phenolic Glycoside from *Psilotum nudum*. Phytochem. **24**, 529 (1985).
83. ACHENBACH, H., and J. WITZKE: Synthese von Psilotin and 6-Epipsilotin. Liebigs Ann. Chem. 2384 (1981).
84. HLUBUCEK, J.R., and A.V. ROBERTSON: (+)-(5)-δ-Lactone of 5-Hydroxy-7-phenylhepta-2,6-dienoic Acid, a Natural Product from *Cryptocarya caloneura* (Scheff.) Kostermans. Austral. J. Chem. **20**, 2199 (1967).
85. SNATZKE, G.: Circular Dichroism and Optical Rotary Dispersion – Principles and Application to the Investigation of the Stereochemistry of Natural Products. Angew. Chem. Internat. Edit. **7**, 14 (1968).
86. BEECHAM, A.F.: The CD of αβ-Unsaturated Lactones. Tetrahedron **28**, 5543 (1972).
87. MEYER, H.H.: Synthesen von (−)-(S)- und (+)-(R)-Goniothalamin; Absolute Konfiguration des natürlichen (+)-Goniothalamins. Liebigs Ann. Chem. 484 (1984).
88. JEWERS, K., J.B. DAVIS, J. DOUGAN, A.H. MANCHANDA, G. BLUNDEN, A. KYI, and S. WETCHAPINAN: Goniothalamin and its Distribution in Four *Goniothalamus* Species. Phytochem. **11**, 2025 (1972).
89. TALAPATRA, S.K., D. BASU, T. DEB, S. GOSWAMI, and B. TALAPATRA: Structure and Stereochemistry of Four New 5,6-Dihydro-2-Pyrones from *Goniothalamus sesquipedalis* and *Goniothalamus grifithii*. Indian. J. Chem. **24B**, 29 (1985).
90. SAM, T.W., C. SEW-YEU, S. MATSJEH, E.K. GAN, D. RAZAK, and A.L. MOHAMED: Goniothalamin Oxide. An Embryotoxic Compound from *Goniothalamus macrophyllus* (Annonaceae). Tetrahedron Letters **28**, 2541 (1987).
91. GOVINDACHARI, T.R., and P.C. PARTHASARATHY: Cryptocaryalactone, A Novel 5,6-Dihydro-2H-pyran-2-one from *Cryptocarya bourdilloni*. Tetrahedron Letters 3401 (1971).
92. SPENCER, G.F., R.E. ENGLAND, and R.B. WOLF: (−)-Cryptocarylalactone and (−)-Deacetylcryptocaryalactone-Germination Inhibitors from *Cryptocarya moschata* seeds. Phytochem. **23**, 2499 (1984).
93. MEYER, H.H.: Synthesen und absolute Konfigurationen von (+)-(6R,2'S)-Cryptocaryalacton und (−)-(6S,2'S)-Epicryptocaryalacton. Liebigs Ann. Chem. 977 (1984).
94. SNATZKE, G., and R. HANSEL: Die Absolutkonfiguration der Kawa-Lactone. Tetrahedron Letters 1797 (1968).
95. ACHENBACH, H., and W. REGEL: Kernresonanzspektroskopische Untersuchungen an Kawa-Lactonen. Chem. Ber. **106**, 2648 (1973).

96. Achenbach, H., and N. Theobald: Notiz zur absoluten Konfiguration der Kawa-Lactone. Chem. Ber. **107**, 735 (1974).
97. Achenbach, H., und G. Wittmann: Dihydrokawain-5-ol, ein neuer Alkohol aus Rauschpfeffer (*Piper methysticum* Forst.). Tetrahedron Letters 3259 (1970).
98. Achenbach, H., and H. Huth: Synthese von Dihydrokawain-5-ol. Tetrahedron Letters 119 (1974).
99. Achenbach, H., W. Karl, and S. Smith: Zur gaschromatographischen Trennung der Kawa-Lactone – (+)-5,6,7,8-Tetrahydro – Yangonin, ein neues Kawa-Lacton aus Rauschpfeffer. Chem. Ber. **104**, 2688 (1971).
100. Achenbach, H., W. Karl, and W. Regel: 11-Hydroxy-12-methoxy-dihydrokawain und 11,12-Dimethoxydihydrokawain, zwei neue Kawa-Lactone aus Rauschpfeffer (*Piper methysticum* Forst.). Chem. Ber. **105**, 2182 (1972).
101. Franca, N.C., O.R. Gottlieb, and A.M. Puentes Saurez: 6-Phenylethyl-5,6-dihydro-2-pyrones from *Aniba gigantifolia*. Phytochem. **12**, 1182 (1973).
102. Matter, U.E., C. Pascual, E. Pretsch, A. Pross, and W. Simon: Estimation of the Chemical Shifts of Olefinic Protons Using Additive Increments. III. Examples of Utility in NMR Studies and the Identification of Some Structural Features Responsible for Deviations from Additivity. Tetrahedron **25**, 2023 (1969).
103. Elvidge, J.A., and P.D. Ralph: Polyene Acids. Part X. The Conformation of Hexenolactone and the Configuration of the Derived Sorbic Acid as Indicated by Proton Magnetic Resonance Spectroscopy. J. Chem. Soc. (B) 243 (1966).
104. Pelter, A., and M.T. Miqdad: The Carbon-13 Nuclear Magnetic Resonance Spectra of Tetronate and 2-Pyrone Derivatives. J. Chem. Soc. Perkin Trans. I 1173 (1981).
105. Urbach, G., W. Stark, and A. Nobuhara: Low Resolution Mass Spectra of Some Unsaturated δ-Lactones. Agr. Biol. Chem. **36**, 1217 (1972).
106. Cardellina, J.H., and J. Meinwald: Isolation of Parasorbic Acid from the Cranberry Plant, *Vaccinium Macrocarpon*. Phytochem. **19**, 2199 (1980).
107. Budzikiewicz, H., C. Djerassi, and D.H. Williams: Mass Spectrometry of Organic Compounds, p. 208. Holden-Day, 1967.
108. Kirk, D.N.: The Chiroptical Properties of Carbonyl Compounds. Tetrahedron **42**, 777 (1986).
109. Lavie, D., I. Kirson, E. Glotter, and G. Snatzke: Conformational Studies on Certain 6-Membered Ring Lactones. Tetrahedron **26**, 2221 (1970).
110. Thomas, S.A.: Conformations of Saturated and Unsaturated δ-Lactone Rings. J. Crystallogr. Spectrosc. Res. **15**, 115 (1985).
111. Clarke, P.J., and P.J. Pauling: Crystal and Molecular Structure of Goniothalamin (+)-(6S)-5,6-Dihydro-6-Styryl-2-Pyrone. J. Chem. Soc. Perkin Trans II 368 (1975).
112. Crombie, L., and P.A. Firth: Biosynthesis of Parasorbic Acid (Hex-2-en-5-olide) by the Rowan Berry (*Sorbus aucuparia* L.). J. Chem. Soc. (C) 2852 (1968)
113. Brereton, R.G., M.J. Garson, and J. Staunton: Biosynthesis of Fungal Metabolites. Asperlactone and its Relationship to Other Metabolites of *Aspergillus melleus*. J. Chem. Soc. Perkin Trans. I 1027 (1984).
114. Leete, E., A. Muir, and G.H.N. Towers: Biosynthesis of Psilotin from [2′,3′-$^{13}C_2$, 1′-^{14}C, 4-^{3}H] Phenylalanine Studied with ^{13}C-NMR. Tetrahedron Letters **23**, 2635 (1982).
115. Haynes, L.J., and E.R.H. Jones: Unsaturated Lactones. Part 1. (Researches on Acetylenic Compounds. Part X.) A New Route to Growth-inhibitory αβ-Ethylenic γ- and δ-Lactones. J. Chem. Soc. 954 (1946).
116. Israili, Z.H., and E.E. Smissman: Synthesis of Kavain, dihydrokavain, and Analogues. J. Organ. Chem. (USA) **42**, 4070 (1976).

Addendum

This addendum includes the literature covered by Chemical Abstracts volume 108 (1988).

RUSSELL, A.T., and G. PROCTER: Allylsilanes in organic synthesis; stereoselective hydroxylactonization of chiral amide-allylsilanes (synthesis of parasorbic acid). Tetrahedron Letters **28**, 2041 (1987).

MURAYAMA, T., T. SUGIYAMA, and K. YAMASHITA: Synthesis of natural (—)-osmundalactone and its epimer. Agric. Biol. Chem. **50**, 2347 (1986).

MURAYAMA, T., T. SUGIYAMA, and K. YAMASHITA: Synthesis of biologically active 6-substituted 5-6-dihydro-5-hydroxy (or acyloxy)-2H-pyran-2-ones. Tennen Yuki Kagobutsu Toronkai Koen Yoshihu 331 (1986).

MURAYAMA, T., T. SUGIYAMA, and K. YAMASHITA: Syntheses of natural (+)-phomalactone, (+)-asperlin and their isomers. Agric. Biol. Chem. **51**, 2055 (1987).

LICHTENTHALER, F.W., K. LORENZ, and W.Y. MA: A convergent total synthesis of (—)-anamarine from D-glucose. Tetrahedron Letters **28**, 47 (1987).

LORENZ, K., and F.W. LICHTENTHALER: A convergent total synthesis of (+)-anamarine from (R,R)-tartrate and D-gulonolactone. Tetrahedron Letters **28**, 6437 (1987).

VALVERDI, S., B. HERRADON, R.M. RABANAL, and M. MARTIN-LOMAS: A synthetic approach to olguine. Can. J. Chem. **65**, 332 (1987).

(Received March 14, 1988)

Building Blocks for the Total Synthesis of Anthracyclinones

By K. KROHN, Institut für Organische Chemie der Technischen Universität, Braunschweig, Federal Republic of Germany

Contents

I. Introduction 37
II. General Synthetic Methods 39
   1. Friedel-Crafts Reactions 39
   2. Diels-Alder Reactions 42
   3. Anionic Reactions 45
   4. Transition Metal Mediated Reactions 48
III. Synthesis of Building Blocks for Rings A and B 49
IV. Construction of Ring A 65
   1. Diels-Alder Reactions 65
   2. Electrophilic Additions 67
   3. Nucleophilic Additions 70
V. Building Blocks for Rings C and D 72
VI. Concluding Remarks 76
References 77

I. Introduction

Anthracyclinones and the glycosidic anthracyclines are pigments, ranging in color from yellow to purple, which are produced by various species of *Streptomyces* (*1, 2, 3*). The very descriptive name *anthracyclines* was coined by BROCKMANN (*1*) and is derived from *anthra*quinones and the likewise tetracyclic tetra*cyclines*. The antibiotic properties of the anthracyclines were observed soon after their discovery (for history see *4*), but the compounds were far too toxic to be useful as antibiotics for treatment of infectious diseases. However, in the late sixties daunorubicin (**1**) was developed by ARCAMONE and his group at Farmitalia as well as MARAL at Rhône-Poulenc (*4*) into an anticancer drug especially successful in the treatment of acute leukemia. Today anthracyclines are among the most often used drugs in antitumor combination chemotherapy and the intensive research on their isolation and structure elucidation, pharmacology, biochemical mode of action as well as chemical synthesis is a result of their importance as antitumor drugs. It is therefore appropriate to demonstrate some of their structural properties on the most important members of the daunorubicin family: daunorubicin (**1**) and doxorubicin (**2**). The somewhat less cardiotoxic semisynthetic 4′-*epi* compounds (**3**) and (**4**) and the corresponding even more effective (but also more toxic) 4-demethoxy compounds (**5**) and (**6**) (idarubicins) obtained by chemical synthesis have also found use in clinical treatment (*4, 5, 6, 7, 8*). Carminomycin (**7**) which was first isolated by Russian authors (*9*) is very toxic and not used as a drug in the Western World.

	R¹	R²	R³	R⁴
1	OMe	H	OH	H
2	OMe	OH	OH	H
3	OMe	H	H	OH
4	OMe	OH	H	OH
5	H	H	OH	H
6	H	OH	OH	H
7	OH	H	OH	H

1 – 7

A large number of anthracyclines have now been isolated and structurally characterized both at academic research centers as well as by chemical companies (*2, 3*). They differ in kind and also in stereochemistry of the substituents of ring A, in the number of sugars (up to eight) attached at C-7, C-9 and C-glycosidically at ring D, and most importantly in the substitution pattern of the anthraquinone nucleus. The substitution pattern that parallels somewhat the chromatographic behaviour is also the basis of the classification and numbering system of BROCKMANN (*1*). In this review we use this convenient system that allows the same numbering for all types of anthracyclines independent of their substitution pattern [see (**1**)]. It must be noted however that the IUPAC system very often requires a different numbering.

Semichemical modifications are much easier to carry out on the side chain or sugar part of the molecule than on the aromatic nucleus.

References, pp. 77–88

Variation of the substitution pattern at the nucleus clearly is within the domain of total synthesis. Previous reviews have emphasized methodological aspects (*4, 5, 6, 10, 11, 12, 13, 14*). This review describes methods for formation of rings A and D and the synthesis of AB building blocks which contain the interesting chiral part of the aglycones and also provides a brief summary of general methods in anthracycline synthesis.

II. General Synthetic Methods

1. Friedel-Crafts Reactions

The structural similarity of anthracyclinones to anthraquinones suggests use of the Friedel-Crafts reaction for the construction of the ring system (*15*). Usually the new bonds are formed between rings B and C as shown in the retrosynthetic Scheme 1 for construction of (**8**) from (**9**) and (**10**). The donor substituted ring in the partially hydrogenated AB building block (**10**) facilitates the electrophilic reaction.

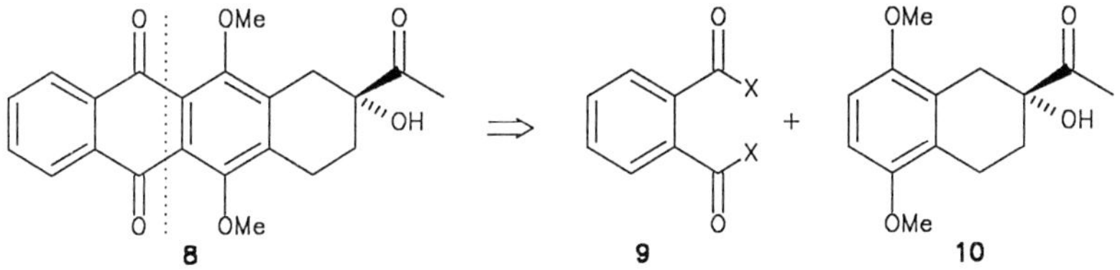

Scheme 1

The new bonds are formed stepwise. The first acylation deactivates ring B of the AB building block so that normally rather harsh reaction conditions have to be used for the second electrophilic attack such as an $AlCl_3/NaCl$ melt at 180 °C (*16*). This procedure is now used at Farmitalia to produce kilogram quantities of 4-demethoxy-7-deoxy-daunomycinone.

Very often this one-pot process is replaced by two distinct steps as realized in the pioneering work of WONG et al. (*17*). The first Friedel-Crafts reaction to yield the ketone (**12**) is effected by treatment of acid (**11**) with the AB building block (**10**) in the presence of trifluoroacetic anhydride. Liquid hydrogen fluoride is then required for the second cyclization step to form (**13a**). Another very important reaction is introduction of the oxygen function at the benzylic position at C-7. This can be done by homolytic bromination to give (**13b**) followed by solvolysis with methanol to (**13c**). The stereoselectivity of this replacement reaction was subsequently improved by means of an elimination-addition reaction to give predominantly the *cis*-diols (*18, 19, 20*) (Scheme 2).

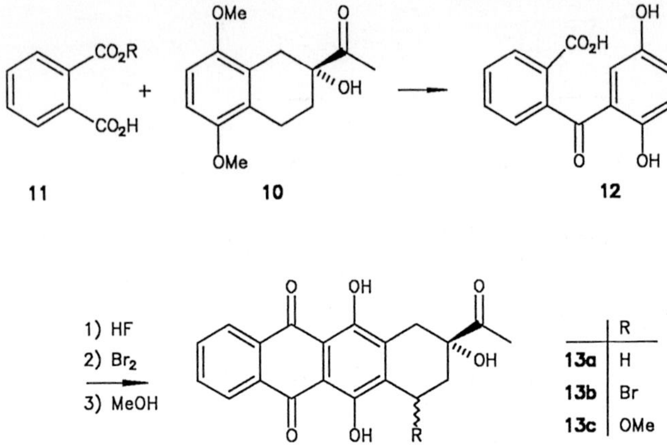

	R
13a	H
13b	Br
13c	OMe

Scheme 2

It is difficult to control regiochemistry with Friedel-Crafts reactions of this type. Hayashi rearrangement may lead to regioisomers even if the first reaction results in site specific coupling of two different building blocks. A very instructive example shown in Scheme 3 is the treatment of benzophenone (**14**) with sulfuric acid which proceeds through the spirocyclic cation (**15**) to afford the rearranged isomer (**16**) (*21*, compare *22*).

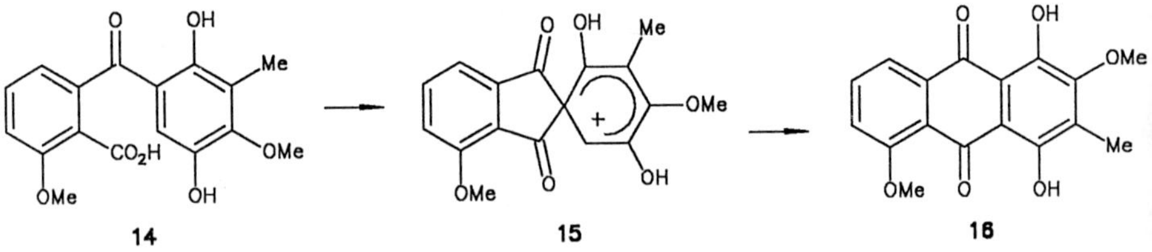

Scheme 3

The lack of regiocontrol can in many instances be overcome by carrying out the reaction in the presence of boric acid as illustrated in Scheme 4 which depicts one step of the aklavinone synthesis of CONFALONE *et al.* (*20*). Formation of an intermediate borate ester (**17**) pre-

References, pp. 77–88

sumably prevents Hayashi rearrangement either by deactivating the phenol or through hydrogen bridging with the neighbouring carbonyl group to give the desired product (**18**) (compare *22*).

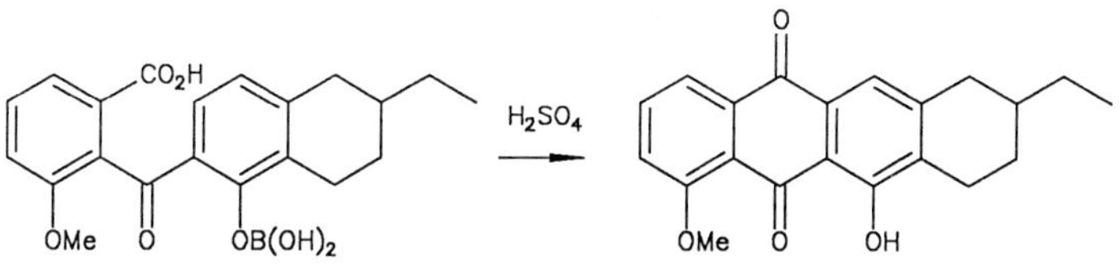

Scheme 4

Regioselective coupling of two unsymmetrically substituted building blocks in the first step of anthracyclinone synthesis is very often achieved by Grignard or Michael reactions. In a daunomycinone synthesis of BRAUN (*23, 24*) the Grignard reagent derived from dihydronaphthalene (**20**) adds regioselectively to 3-methoxyphthalic anhydride (**19**) to afford (**21**) (Scheme 5). Cyclization to the tetracycle (**22**) is smoothly effected by treatment with liquid hydrogen fluoride [for rationalization of regiochemistry see (*25*)].

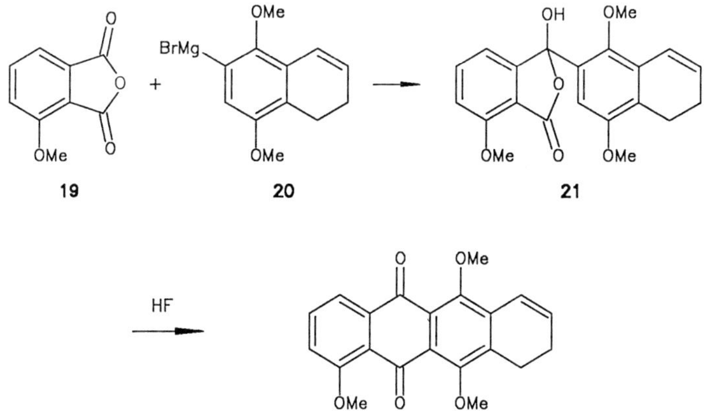

Scheme 5

In the examples discussed so far both quinone carbonyls are derived from one building block in accordance with the cyclization mode shown in Scheme 1. A Michael addition of cyanide (**23**) to the unsaturated ester (**24**) affords adduct (**25**) used in the 7,9-dideoxy daunomycinone synthesis of PARKER and KALLMERTEN (26) (Scheme 6). The subsequent Friedel-Crafts reaction of the corresponding acid with trifluoroacetic acid and trifluoroacetic anhydride gives the tetracyclic compound (26). The dotted line in (26) illustrates the origin of the carbon atoms (for similar Michael additions compare 27, 28).

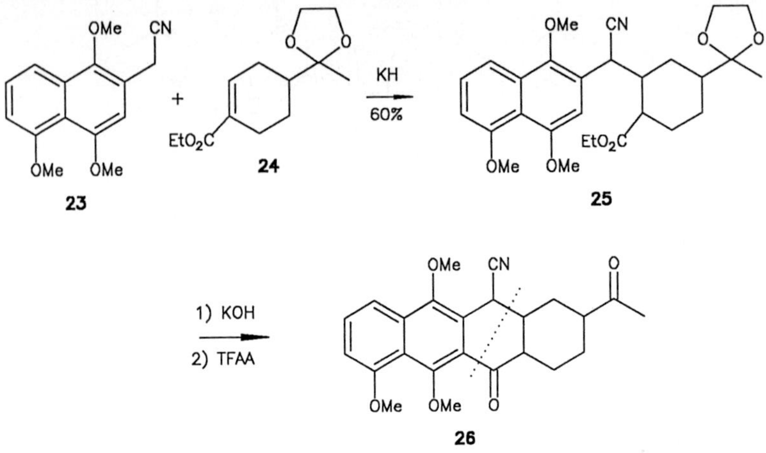

Scheme 6

The Friedel-Crafts reactions in anthracyclinone synthesis are limited to the coupling of partially functionalized building blocks due to the possibility of elimination under the strongly acidic reaction conditions. Furthermore only anthracyclines of the daunomycinone family with an acetyl side chain can be synthesized using this method. Presence of a carbonyl group in the acetyl side chain inhibits development of a partial positive charge on the neighbouring tertiary carbon atom and this in turn prevents the acid catalyzed dehydration that occurs very easily with rhodomycinones having an alkyl side chain. Another difficulty with the Friedel-Crafts reaction is racemization during the $AlCl_3/NaCl$ melt at 180 °C. Careful studies by several authors have shown that at least partial racemization does occur during this reaction (29, 30).

2. Diels-Alder Reactions

[2+4] Cycloadditions offer ideal possibilities for the construction of linearly condensed ring systems containing sixmembered rings. In fact, six of the eight possible ring couplings shown in (27) have been realized (disconnetions a–f). The Diels-Alder reaction allows the construction of up to eight stereocenters in one step (*31, 32*). However, most of the chiral centers in anthracyclines are destroyed in subsequent aromatization reactions.

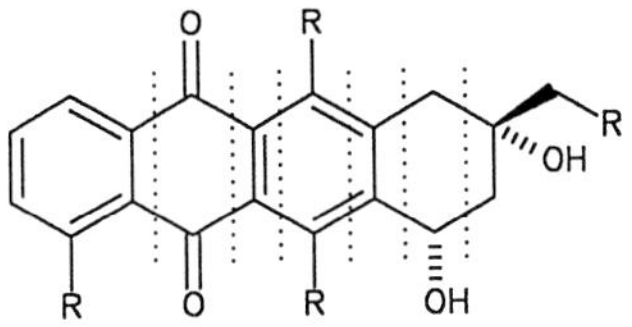

27

Control of regiochemistry is of great importance. Regiocontrol is more difficult to achieve in rhodomycinones derived from 1,4,5-trihydroxyanthraquinone than in 11-deoxy compounds such as the aklavinones (**36**). This is nicely illustrated by the successful regiocontrol achieved by TAMIREZ and VOGEL (*33*) in the cycloaddition of the intramolecularly linked acrylate (**28**) to the adduct (**29**) (Scheme 7). This reaction is reminiscent of a method described by BRESLOW *et al.* (*34*) for controlling site specific reactions in steroids.

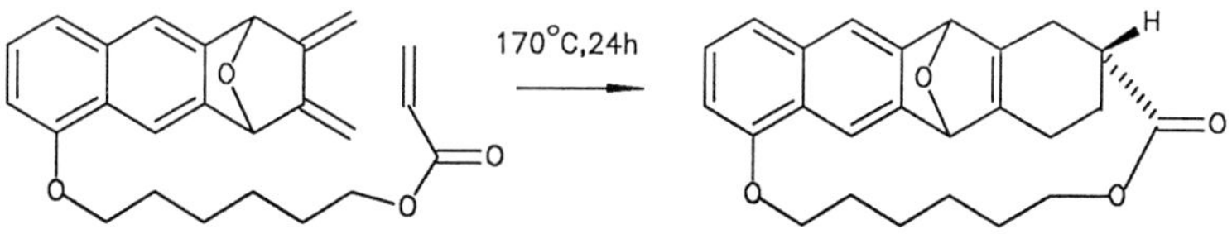

Scheme 7

The regioselectivity of the Diels-Alder reaction of naphthoquinones is strongly influenced by the presence of electron-donating or withdrawing substituents. A strongly chelated *peri* hydroxyl group on C-5 which withdraws electron density from the neighbouring carbonyl group is often able to direct the regiochemical outcome. Thus the partially acylated naphthazarin derivative (**30b**) which is in equilibrium with (**30a**) adds to bistrimethylsiloxy diene (**31**) to afford the adduct (**32**) as a single regioisomer (*35*) (Scheme 8). This principle of two successive Diels-Alder additions to naphthazarin which results in the formation of the future rings B and C is very often used in the synthesis of daunomycinone derivatives (*36, 37, 38*).

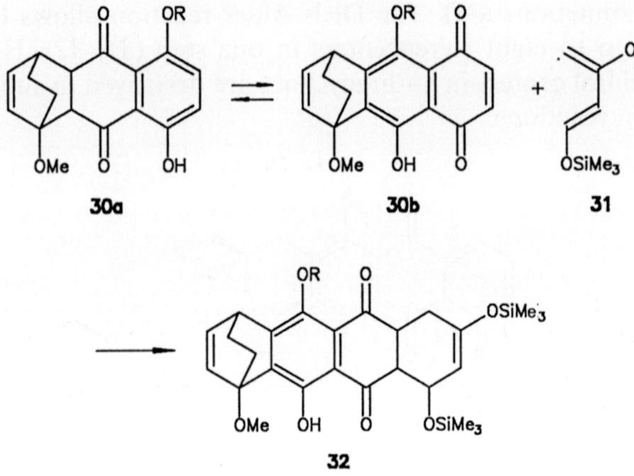

30a **30b** **31**

32

R = p—nitrobenzyloxycarbonyl

Scheme 8

An example of the many reactions which allow good regiocontrol is shown in Scheme 9. Reaction between juglone (**33**) and alkoxybutadienes such as (**34**) yields adduct (**35**) (*39*), which is further transformed to akalavinone (**36**) (*40*).

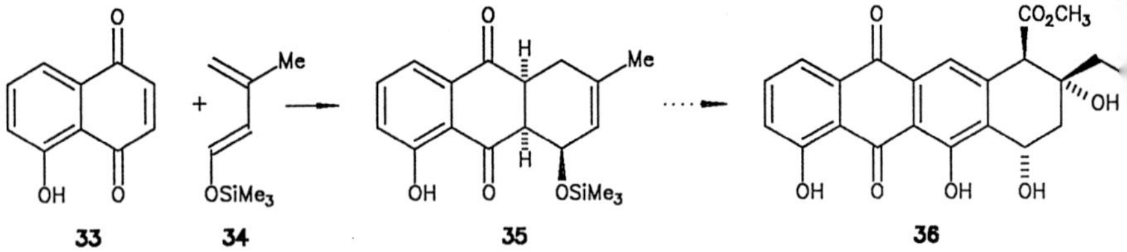

33 **34** **35** **36**

Scheme 9

Diels-Alder reactions of the methyl ether of juglone (**33**) normally afford adducts of inverse regiochemistry (*39, 41, 42*). This can be rationalized in terms of frontier orbital interactions (*43*). However, halogen atoms (notably bromine) on the quinone nucleus are also very effective in controlling the regiochemical outcome (*44, 45, 46*). In a very detailed study GESSON et al. (*46*) investigated the reaction of naphthoquinones with a variety of open chain and exocyclic dienes. In the reaction of the bromojuglone derivate (**37**) with the ketene acetal (**38**) adduct (**40**) is eventually formed overcoming the inverse directing properties of the methoxy group in (**37**). An ionic mechanism *via* the intermediate (**39**) has been postulated to explain this behaviour (*46*) (Scheme 10).

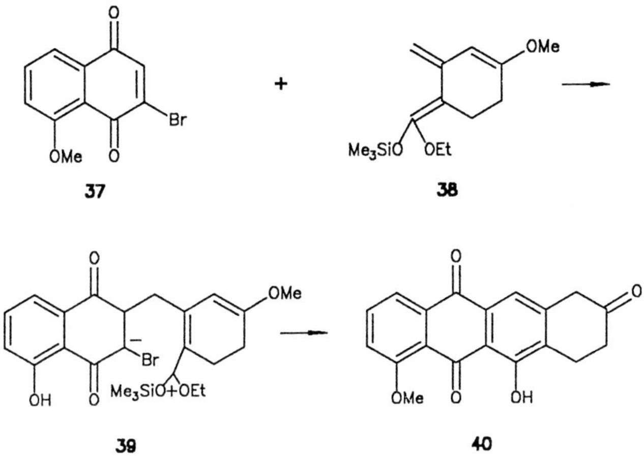

Scheme 10

3. Anionic Reactions

Anionic reactions of two appropriate building blocks exert good control over both regio- and stereochemistry. Reports describing the formation of rings A, B, and C by addition of organometallic reagents to carbonyl compounds by Michael additions, variations of aldol reactions and electrophilic as well as nucleophilic additions to aromatic systems have been published.

Almost ideal starting materials for the synthesis of tetracyclic anthracyclinones are the readily available and cheap hydroxylated anthraquinones. A very straightforward procedure originally discovered by MARSCHALK et al. in 1936 (47) can be employed to overcome the poor reactivity of the electron deficient anthraquinone system (Scheme 11). A hydroxylated anthrahydroquinone obtained upon dithionite reduction reacts with aldehydes at 100 °C to form alkylated anthraquinones. The initially formed benzylic hydroxy group is eliminated at the high reaction temperatures in an intramolecular redox process. This synthetically valuable hydroxy group can however be preserved by lowering the reaction temperature to 0–5 °C (48, 49). The Marschalk reaction has been used twice in a total synthesis of rhodomycinone (**44**) as shown in the hydroxymethylation of 1-hydroxy-4,5-dimethoxy-9,10-anthraquinone (**41**) to (**42**). The second intramolecular Marschalk reaction of the intermediate hydroxyaldehyde (**43**) directly gives γ-rhodomycinone (**44**) (18, 50). Conditions were found under which the desired trans 1,2-diol was obtained almost exclusively. Also regiochemistry can

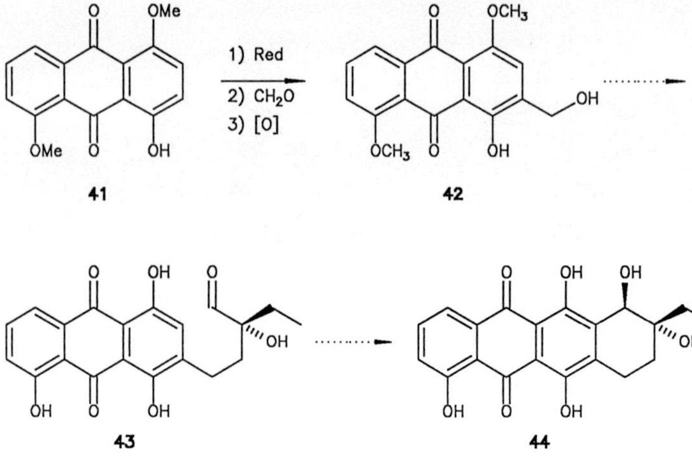

Scheme 11

easily be controlled by selective protection (*18, 50*) or by employing different reaction conditions during the first alkylation step (*51*).

The procedure using the Marschalk reaction of hydroxyanthraquinones is at present the best method for the synthesis of enantiomerically pure rhodomycinones. Easily available chiral building blocks derived from sugars (*52, 53, 54*), lactic acid (*55*) or α-aminobutyric acid (*56*) can be incorporated. The alkylation of anthraquinones can also be effected by reductive Claisen rearrangement (*57*) or even by nucleophilic addition to anthraquinones. Nitronates have been shown to be particularly useful building blocks due to elimination of the nitro group by a redox process similar to elimination of the benzylic hydroxy group in the Marschalk reaction. The reaction is illustrated by the cyclization of (**45**) to the carminomycinone precursor (**46**) (*58, 59*) (Scheme 12).

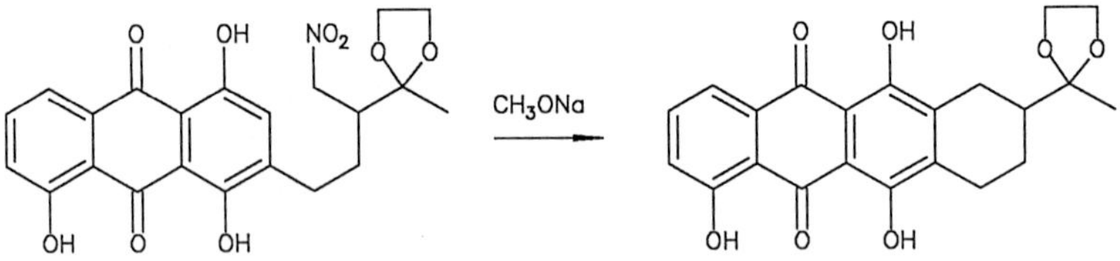

Scheme 12

An aldol reaction is a key step in biogenetic type syntheses of anthracyclinones with an ester group at C-7 [referred to as type B (*60*)] (Scheme 13). The first total synthesis of ζ-rhodomycinone (**48**) was accomplished in this manner by cyclization of the keto ester (**47**) (*61*). The side chains in (**47**) can be introduced by two successive Marschalk reactions starting from anthraquinone (**41**). This aldol cyclization is the method of choice in type B anthracyclinone syntheses irrespective of the substitution pattern of rings B or D. The doubly activated benzylic position adjacent to the ester group always reacts as the nucleophile. The stereochemistry of the newly created stereogenic centers (*62, 63, 64, 65, 66*) as well as the absolute configuration can be controlled (*45*).

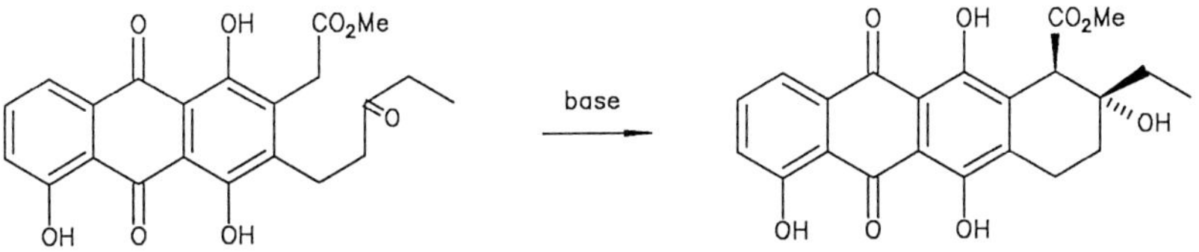

Scheme 13

The Grignard (*23*) and Michael reactions (*26, 27, 28*) have already been mentioned in connection with subsequent Friedel-Crafts cyclizations (Scheme 5). Another illustrative example is the coupling of the doubly lithiated benzamide (**49**) with aldehyde (**50**) to afford the 11-deoxydaunomycinone intermediate (**51**) (*67*) (Scheme 14). A similar strategy was successfully applied in the synthesis of aklavinone (*68*).

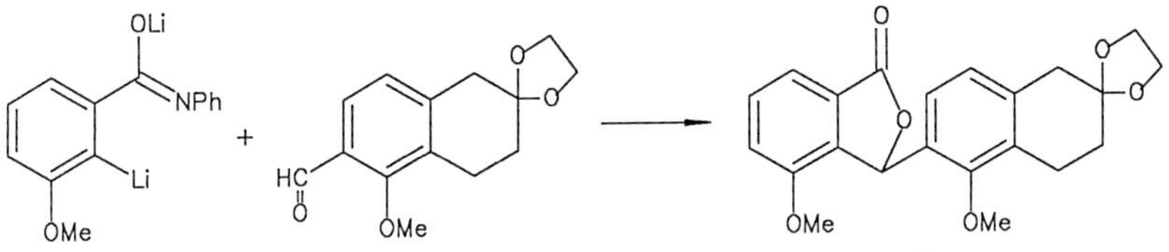

Scheme 14

A remarkably selective and convergent approach is the coupling of phthalides which act as 1,4-dipoles with various kinds of Michael acceptors (*69, 70, 71, 72, 73*). The example in Scheme 15 taken from the daunomycinone synthesis of SWENTON *et al.* (*73*) illustrates the convergency of the one step coupling of (**52**) and (**53**) to the precursor (**54**). The mild reaction conditions are compatible with a high degree of functionality in the AB building block (**53**).

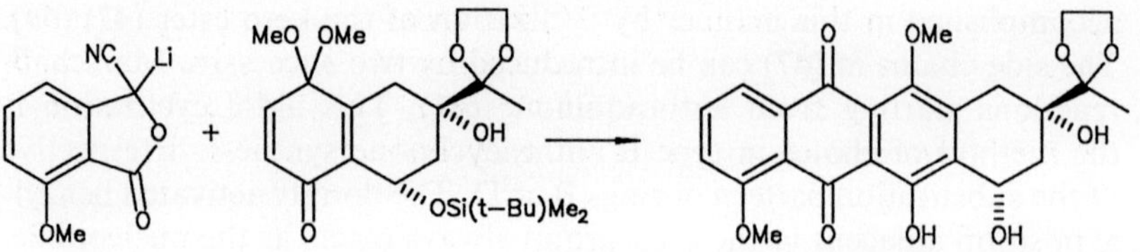

48   K. Krohn:

Scheme 15

4. Transition Metal Mediated Reactions

Transition metals play an increasing role in organic synthesis. This is also true for anthracyclinone chemistry. Regioselective construction of the carbon skeleton and chemoselective reactions have been achieved using chromium carbonyl complexes. Two different routes have been explored. In the first approach described by KÜNDIG *et al.* (*74*) the complex serves to modify the chemical reactivity. One example is the regioselective addition of carbanions such as 2-lithio-2-methyl-1,3-dithian (**56**) to afford adduct (**57**) on reaction with complex (**55**) (Scheme 16).

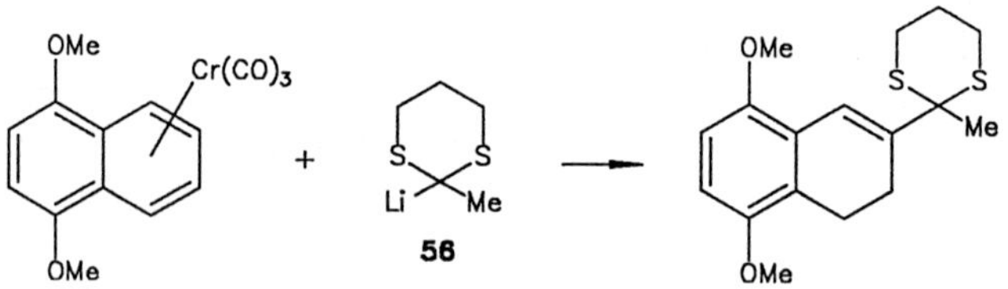

Scheme 16

The second approach makes use of transition metals in the construction of new hydroquinone rings through cycloaddition reaction of chromium complex (**58**) with acetylenes such as (**59**) to give the anthracyclin-

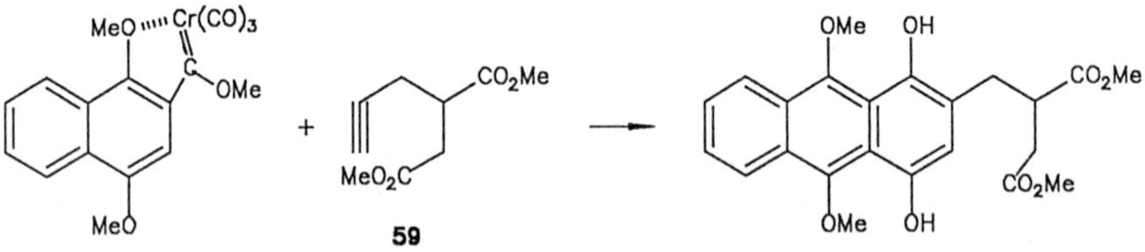

Scheme 17

References, pp. 77–88

one precursor (**60**) (Scheme 17) (*75*, for general methodology see *76*). Other reactions of this type have been studied in the synthesis of daunomycinone precursors by the group of DÖTZ (*75, 77, 78*), WULFF (*79, 80*) and in approaches to nogalamycin by SEMMELHACK (*81*).

III. Synthesis of Building Blocks for Rings A and B

An important strategy for the synthesis of anthracyclinones involves coupling of AB building blocks in a convergent manner to afford the tetracyclic skeleton. This applies to all methods mentioned in Chapter II. However, the functionalities which can be tolerated in the building blocks depend on the coupling method and the elegance of the synthesis depends on the number of steps necessary to adjust functional groups and stereochemistry. In the following synthesis of highly functionalized fragments, particularly in enantiomerically pure form, will therefore be emphasized.

Methods for the synthesis of AB fragments can be divided into those which assemble side chain and functional groups around a benzene nucleus by stepwise cyclization and those that directly start with bicyclic derivatives such as α- or β-tetralones. The Diels-Alder reaction is also very often used to construct the partially hydrogenated naphthalene derivatives.

The first AB building block suitable for coupling with phthalic acid derivatives to give daunomycinones was prepared by WONG et al. (*17*). After condensation of 1,4-dimethoxy-benzaldehyde (**61**) with acetylacetone, the product is hydrogenated to (**62**) and alkylated with bromoacetic ester. The correspondig acid is best cyclized with liquid hydrogen fluoride to afford diketone (**63**) which can be chemoselectively hydrogenated to (**64**). Reaction at the acidic position with oxygen and reduction of the intermediate hydroperoxide with zinc affords the α-hydroxy ketone (**10**) (Scheme 18).

A further improvement introduced by WONG et al. (*82*) is alkylation of (**62**) with allyl bromide to afford after *retro*-Claisen reaction the olefin (**65**). The aldehyde (**66**) obtained upon ozonolysis of (**65**) can be cyclized with hydrogen chloride. Reduction to (**64**) can be achieved either with triethylsilane (*82*) or by catalytic hydrogenation (*83*) (Scheme 19). Acetyltetralins such as (**63**) and (**64**) are very often used as intermediates in daunomycinone synthesis; resolution of racemates can be effected by using chiral Schiff bases derived from (**10**) (*73*).

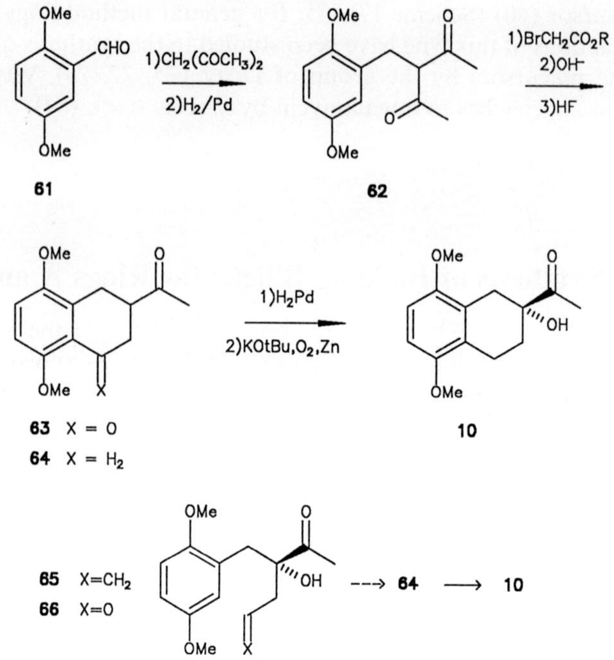

Scheme 18

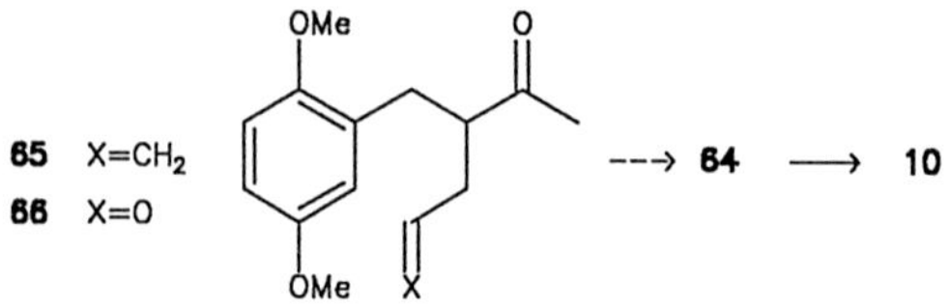

Scheme 19

Variations of this strategy have been introduced by RAO and co-workers (*84*). Selective cleavage of the neighbouring methyl ether in (**63**) with boron trifluoroetherate yields the chelated phenol (**67**) which can in turn be used in a regioselective synthesis of daunomycinone (*85*). A corresponding monophenol (**69**) that is useful for 11-deoxy-daunomycinones (*86*) can be obtained if the bromoanisole (**68**) is used as starting material (Scheme 20).

Similar Friedel-Crafts reactions can be employed in the synthesis of highly functionalized enantiomerically pure building blocks. The starting material in the work of GENOT, FLORENT and MONNERET (*87*), α-D-isosaccharino-1,4-lactone (**70**), can be converted to the aldehydes (**71**) and (**75**). These aldehydes react with lithiodimethoxybenzene (**72**) to give adducts that can be transformed to the aldehydes (**73**) and (**77**). Cyclization with tin tetrachloride affords the optically active AB building blocks (**74**) and (**78**) (Scheme 21).

A very short reaction sequence involving incorporation of malic acid (**79**) which leads to the fully functionalized optically active building block (**85**) (*88*) is shown in Scheme 22. Direct α-alkylation of α-alkoxy esters would lead to complete racemization due to enolization. This difficulty was overcome by SEEBACH *et al.* (*89*) by acetalization of malic acid with pivalic aldehyde to yield predominantly the *cis* product (**80**) in a kinetically controlled reaction. Alkylation of (**80**) with 2-iodo-1,4-dimethoxybenzene (**81**) leads to (**82**) with high diastereoselectivity

(>97% de). The corresponding acid chloride of (**82**) is cyclized with tin tetrachloride to the tetralone (**83**) followed by zinc borohydride reduction and chain elongation [NaCH$_2$SOCH$_3$, Al/Hg (*90*)]. Equilibration of the epimeric diols can be effected with phenylboronic acid (*91*) to yield (**84**) in which the hydroxy groups are already protected for the subsequent oxidation with ceric ammonium nitrate (CAN) which affords the quinone (**85**) (*88*) (Scheme 22). This is perhaps the shortest synthesis to-date of an enantiomerically pure AB building block for daunomycinone synthesis. Quinones such as (**85**) react with diketenes formed by photolysis of benzocyclobutenediones to give symmetrically substituted daunomycinones (*92*) (see Scheme 59).

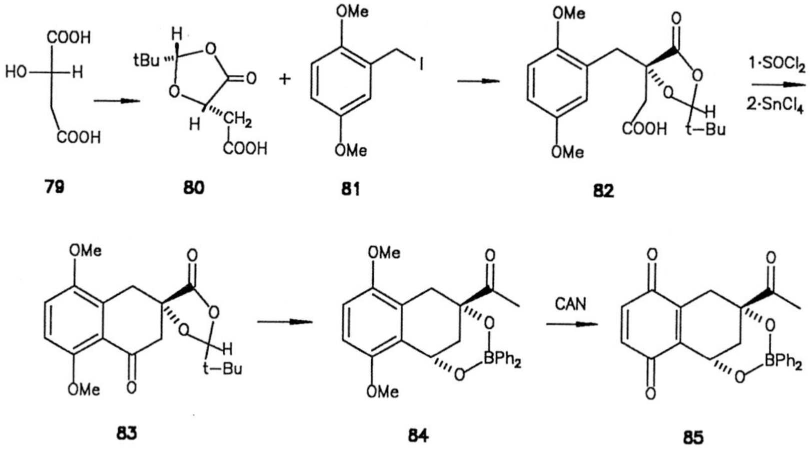

Scheme 22

Other important key intermediates introduced by WONG *et al.* (*93*) were the acids (**87**) and (**88**) obtained by HF treatment of the benzylated succinic acid anhydride (**86**) (Scheme 23). Conversion of the carboxyl group to an acetyl side chain by reaction of (**88**) with methyl lithium affords (**64**). This reaction is often used in anthracyclinone chemistry (*94, 95*). Nucleophilic addition of sodium dimethylsulfoxide followed by reduction with aluminum amalgam is an alternative method for chain elongation of esters or lactones introduced by COREY *et al.* (*90*) [see (**83**)→(**84**)] (for application of this method see *73, 88, 91, 96*).

The carboxylic acids (**87**) and (**88**) have been used as intermediates in the synthesis of highly functionalized AB building blocks such as (**89**) and (**90**) (*97, 98*) (Scheme 24). Since the benzylic methoxy groups have to be cleaved in the eventual tetracyclic products, a further im-

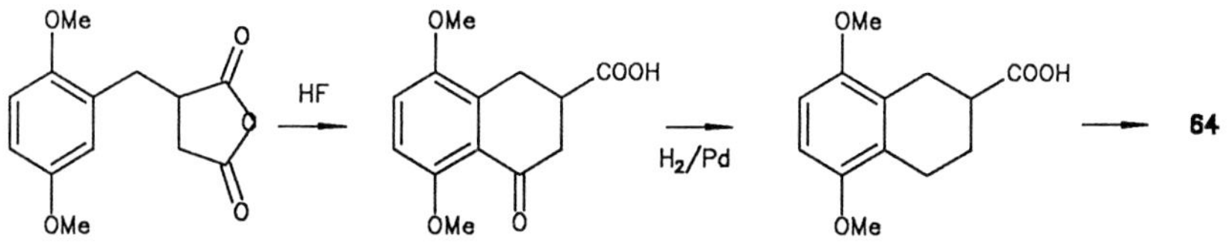

Scheme 23

provement was introduced with the synthesis of the corresponding silyl ether (**94**) with correct stereochemistry. To this end esterification of acid (**87**), thioacetalization and chain elongation using the method of COREY (*90*) affords compound (**91**) in which the two carbonyl groups are chemically differentiated. Hydroxylation of the acidic position by the method of GARDNER (*99*) [(KOt-Bu; O_2, $P(OR)_3$], and resolution by means of a Schiff base with (−)-α-methylbenzyl amine (compare *100*) gives an optically active α-hydroxy ketone that is then ketalized to (**92**). An important step is the stereoselective reduction of the ketone obtained after dethioacetalization of (**92**) with potassium selectride to the *cis*-diol (**93**). Finally, deacetalization, silylation, electrochemical oxidation to the benzoquinone bisacetal and regioselective (ratio: 85:15) cleavage yields the highly functionalized AB building block (**94**) (Scheme 24).

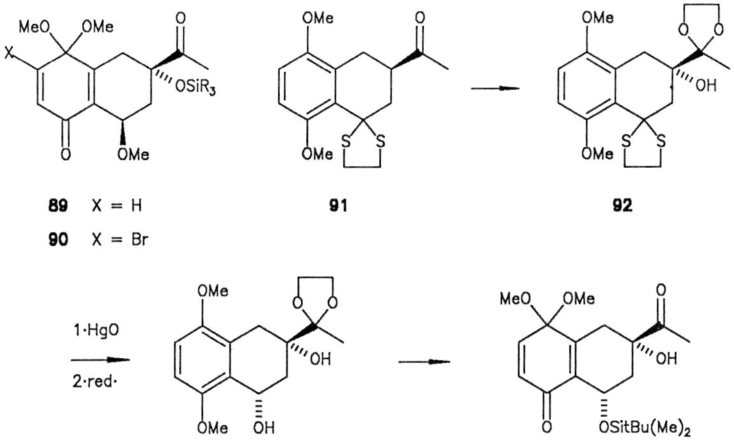

Scheme 24

Two modifications of this sequence were introduced by BROAD-HURST, HASSALL and THOMAS (*96, 101*). Resolution can be achieved at the carboxylic acid stage and equilibration of the *cis/trans* mixture by treatment with phenylboronic acid then gives cyclic esters similar to (**84**). Not only the acetyl side chain but a variety of other groups such as hydroxymethyl, hydroxyethyl, cyanomethyl or methyl and ethyl can be obtained from the carboxylic acid (**88**) (*102*).

Unsaturated carboxylic acids similar to (**88**) have served in two other ways in the preparation of enantiomerically pure AB building blocks. The bromolactonization procedure developed by the group of TERASHIMA (*103*) proceeds *via* intermediates (**95**) and (**96**) and gives the optically active α-hydroxy ketone (**10**). A direct chain elongation resulting in an acetyl side chain can be achieved starting from an amide intermediate derived from (**96**) (*104*) (Scheme 25).

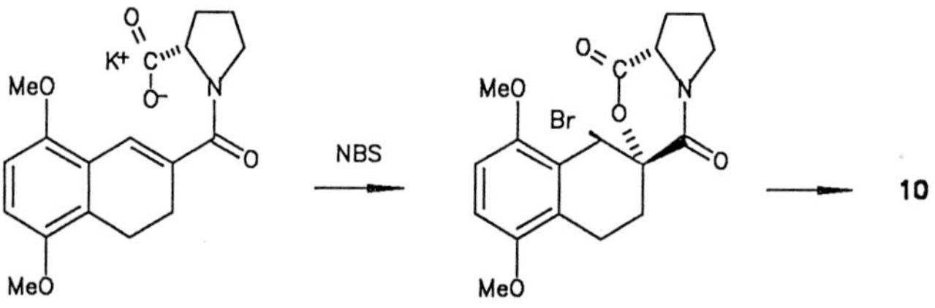

Scheme 25

The enantioselective *cis*-hydroxylation of unsaturated acids such as (**97**) with osmium tetroxide in the presence of chiral bases such as diamine (**98**) is of general importance. Reduction of the product (**99**) to the optically active α-hydroxy ester (**100**) can be achieved with triethylsilane (*105*) (Scheme 26).

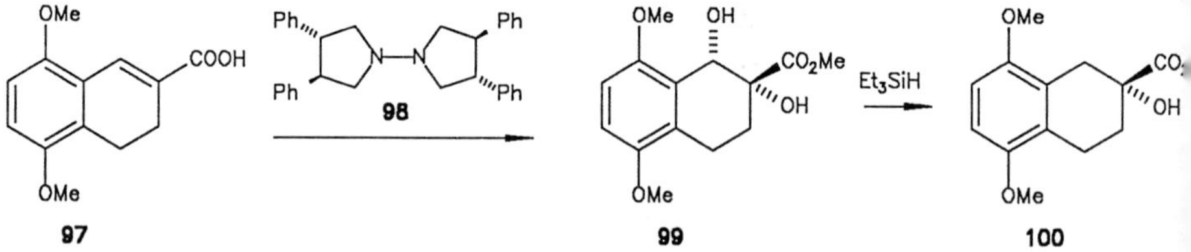

Scheme 26

α-Tetralones are readily available (*106*) and are very often used as starting materials for AB building blocks. The following examples illustrate how the β-side chain and the tertiary hydroxy group are introduced. An important feature is the selective cleavage of the methyl ether adjacent to the carbonyl group as shown in the conversion of (**101**) to (**102**) (Scheme 27). Ether cleavage is effected by potassium iodide in formic acid (*104*), boron tribromide (*107*) or boron trifluoride etherate (*85*). The chemical differentiation introduced thereby can be used in directing the oxidation of the corresponding hydroquinone monomethyl ethers to quinone monoacetals such as (**104**). The side chain of (**103**) is introduced by an ester condensation (compare *100*, *108*) with formic ester. Enantioselective hydroxylation at the tertiary position can be achieved by means of the chiral auxiliary proline (*104*) [(compare conversion of (**95**) to (**96**) (*103*)].

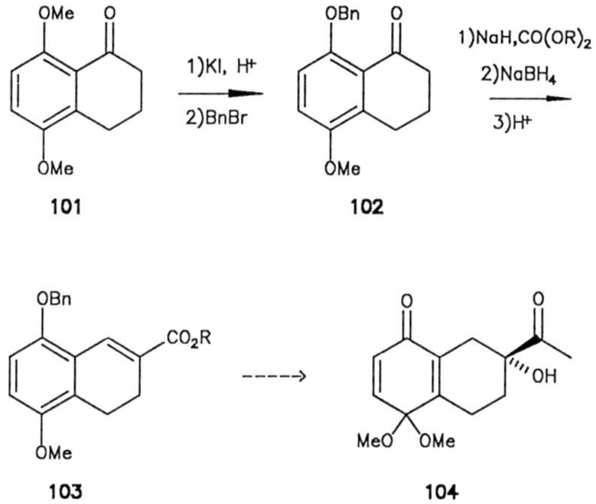

Scheme 27

The chemical differentiation of the two oxygen functions was used in yet another way (*107*). Formation of a mono-methoxymethyl ether (MOM ether) (**105**) activated the neighbouring position for metallation, thus permitting regiospecific introduction of a formyl group to afford the aldehyde (**106**) (Scheme 28). An example which shows use of a similar aldehyde (**50**) in a coupling reaction is shown in Scheme 14.

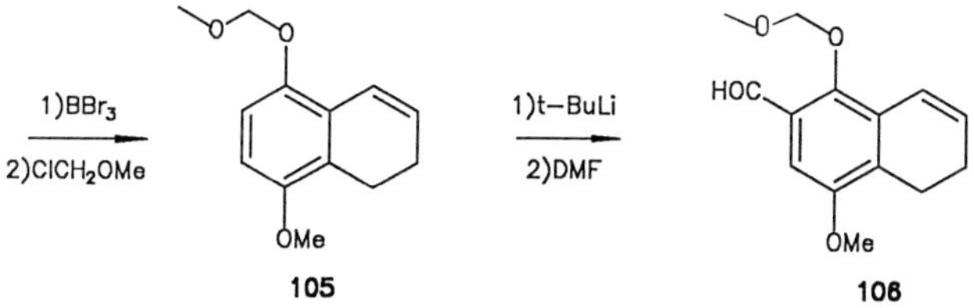

Scheme 28

Side chains can be introduced not only by ester condensations with α-tetralones but also by acylation of olefins such as (**107**) which are available from α-tetralone (**101**) (*110*) (Scheme 29). Using a different protection deprotection strategy the regioisomer (**109**) of (**104**) can be obtained from the acylated product (**108**) (*110*). An alternative to the acylation reaction is the stepwise introduction of a formyl group by a Vilsmeyer reaction followed by reaction with methylmagnesium iodide (*111, 112, 113*).

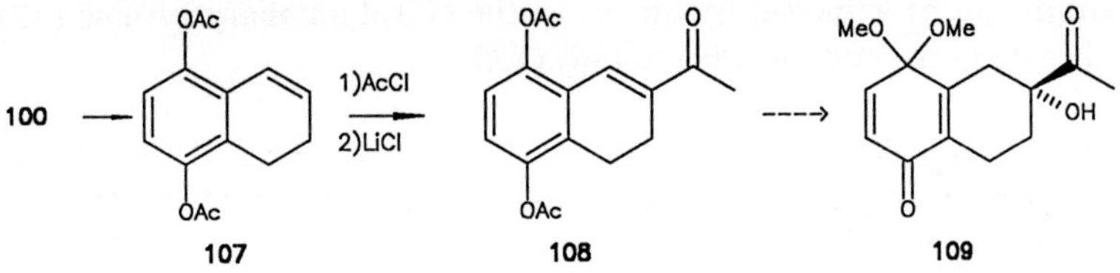

Scheme 29

Fries rearrangements also offer a short way of introducing the requisite functional groups. Thus the acetylnaphthalene (**111**) can be obtained from (**110**) in just two steps (*114*). A special feature of 2-acetylnaphthalene (**111**) is the selective Birch reduction of the acylated ring with simultaneous elimination of the methoxy group to yield (**112**) (Scheme 30). The latter can be transformed further to the AB building block (**113**) which is useful in the synthesis of 11-deoxydaunomycinones (*115*). A related sequence also leads to the daunomycinone precursor (**10**) (*116*).

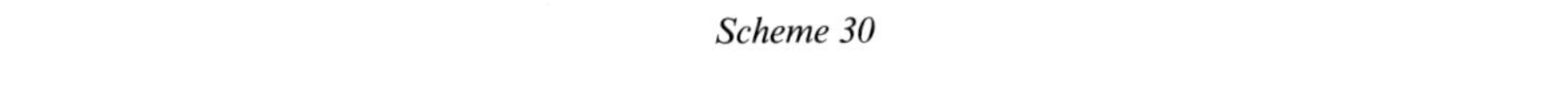

Scheme 30

Another means of introducing functionalized side chains into aromatic systems is the Claisen rearrangement of allyl ethers such as (**114**) which can also be used in corresponding tricyclic hydroanthraquinones (reductive Claisen rearrangement) (*57, 117, 118*). The double bond of the rearranged Claisen product can be isomerized to the thermodynamically stable conjugated position and then cleaved by ozonolysis. Methy-

References, pp. 77–88

lation of the intermediate aldehyde affords the AB fragment (**50**) which serves as a building block for 11-deoxydaunomycinone (*67*) (Scheme 31).

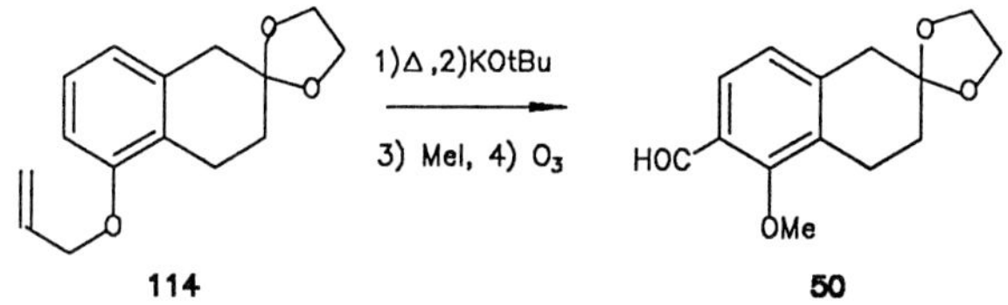

Scheme 31

For construction of anthracyclinones containing an ethyl side chain it is advisable to start with α-tetralones which already incorporate the side chain. A typical example from the work of BRAUN (*24*) shows cyclization of brominated acid (**115**) to the intermediate ethyl tetralone (**116**) which can undergo reduction and subsequent elimination to afford the olefin (**117**) (Scheme 32). Important in this connection is the observation (**109**) that two hydroxyl groups may be introduced simultaneously into (**116**) upon oxygenation employing the method of GARDNER (*99*). The oxygenated product (**118**) can be further transformed to the highly functionalized building block (**119**), a precursor for the 6-deoxyanthracycline β-citromycinone (*120*) (Scheme 32).

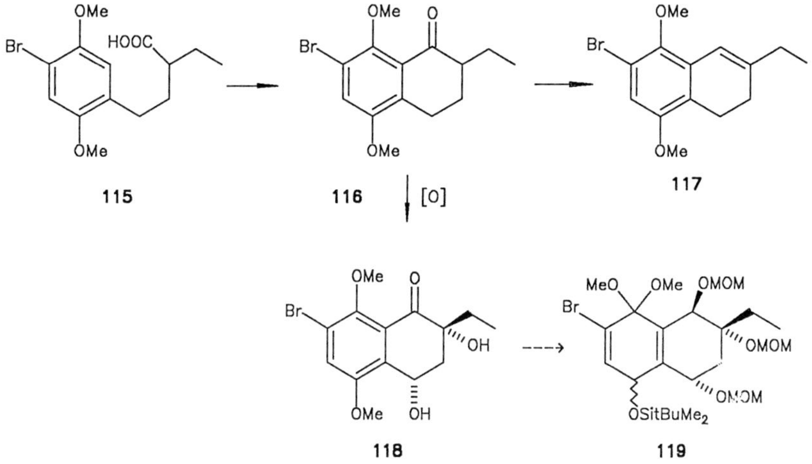

Scheme 32

KENDE and RIZZI have transformed the monomethoxy analogue (**120**) of (**116**) into an advanced intermediate for their aklavinone synthesis (*68*). One interesting step is the quenching of the intermediate vinylic anion generated with strong base from the sulfonylhydrazone (**121**) with formaldehyde or DMF to afford the hydroxymethylated olefin (**122**) (Scheme 33). A second formyl group is introduced by direct metallation of the aromatic ring in the position *ortho* to the methyl ether and quenching with DMF to (**123**). The allyl alcohol generated from the silyl ether (**123**) served as a substrate for enantioselective SHARPLESS epoxidation of a tetracyclic intermediate.

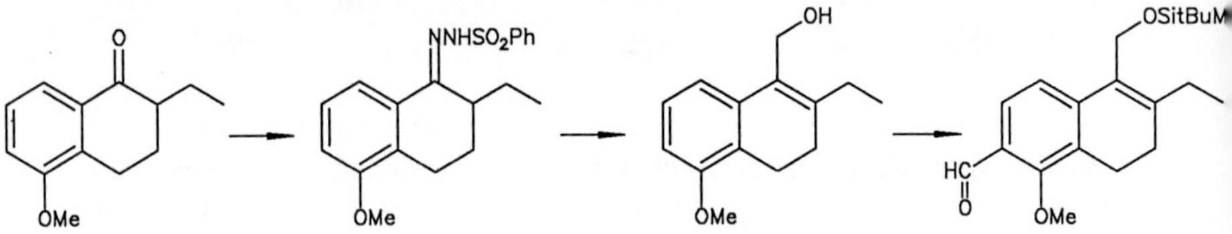

Scheme 33

A building block with ethyl side chain for an aklavinone synthesis was constructed in a Diels-Alder reaction by LI *et al.* (*121*) from benzoquinone (**124**) and the diene (**125**) (Scheme 34). Both reaction partners are relatively electron deficient and therefore not very reactive. Condensation at higher reaction temperatures led to almost complete aromatization of the primary adduct (**126**). The problem was solved by carrying out the reaction at high pressure (15 kbar). The adduct (**126**) was obtained in high yield and directly reduced by zincborohydride or Meerwein-Ponndorf-Verley reduction to the enone (**127**), an ideal reaction partner for Michael additions (*121*).

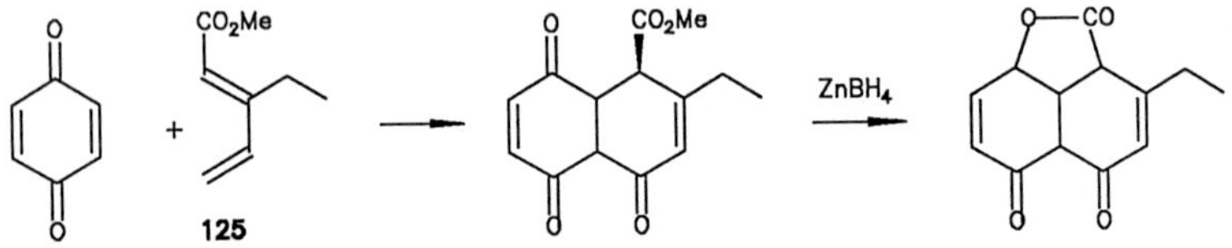

Scheme 34

A similar enone was synthesized by HAUSER and MAL (*122*) in a completely different way starting from the bicyclic acetate (**128**) (Scheme 35). Cleavage of the double bond with ruthenium tetroxide/sodium periodate gave a diacid which in the form of its diacid chloride was converted to the diketone (**129**) by treatment with methyl copper.

References, pp. 77–88

Saponification, oxidation, lactonization and Claisen condensation afforded (**130**) which was coupled to a 11-deoxydaunomycinone precursor using the sulphonylphthalide procedure similar to the reaction outlined in Scheme 15.

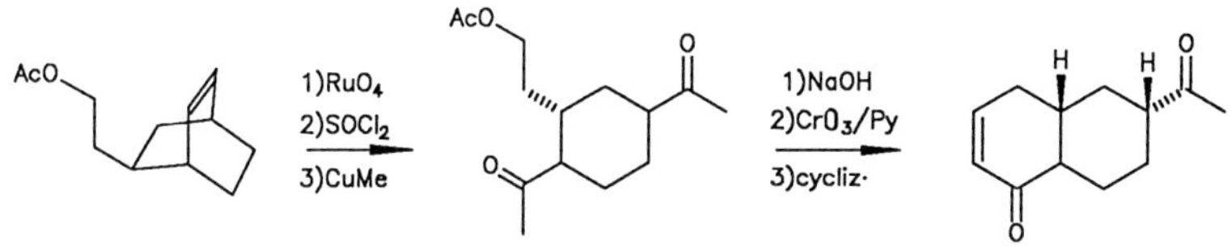

128 **129** **130**

Scheme 35

β-Tetralone derivatives are also very useful precursors for AB building blocks due to the ease with which the tertiary hydroxy group can be generated by addition of organometallic reagents. Dihydronaphthalene intermediates such as (**132a**)–(**132c**) are very often synthesized by Diels-Alder reaction of benzoquinone (**124**) with butadienes such as (**131a**)–(**131c**). The oxygen function is introduced by hydroboration of adduct (**132a**), whereas acid treatment of intermediates (**132b**) and (**132c**) from 2-methoxybutadiene (*123*) or chloroprene (*108*) directly leads to β-tetralone (**133**) (Scheme 36). The requisite side chains can now be introduced to afford the important AB building block (**10**) by reaction of (**133**) with a variety of organometallic reagents such as ethynyl magnesium bromide (*108, 124, 125*), hydrogen cyanide followed by THP protection and reaction with methyl magnesium iodide (*126, 127*), lithio methyl vinyl ether (*128, 129*), phosphorous ylide like $PH_3P{=}CHCO_2R$ (*130*), trimethylsilylethynyl cerium reagents such as $CeCl_2C{=}CSiMe_3$ or $Ce(C{\equiv}CSiMe_3)_3$ (*131, 132, 133*) (conversion of tetracycles with cerium reagents (*134, 135*)], allyl manganeous species generated from allylbromide and a mixture of manganese(II) chloride and lithium aluminum hydride [it is claimed that this reagent reacts even with easily enolizable ketones (*136, 137*)], and finally lithio 2-methyl-1,3-dithiane (*123, 127*).

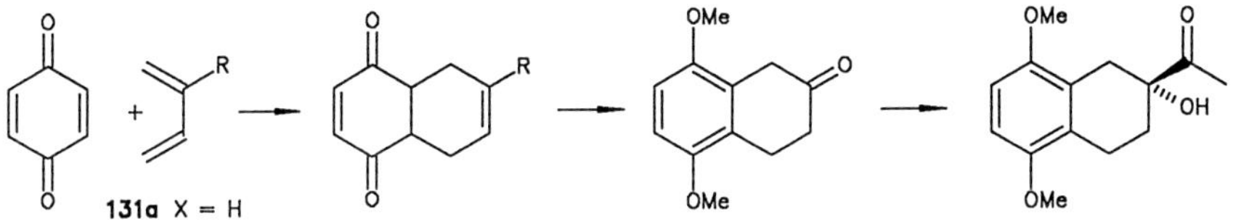

124 **131a** X = H **131b** X = OMe **131c** X = Cl

132a X = H **132b** X = OMe **132c** X = Cl

133 **10**

Scheme 36

Cyanide addition to β-tetralone (**133**) can be used to synthesize 9-aminodaunomycinones by way of the α-aminoketone (**135**) (*138*). The amino ketone is obtained from (**133**) by means of a Strecker type amino acid synthesis to (**134**) followed by resolution with mandelic acid and chain elongation using the method of COREY (*90*) (Scheme 37).

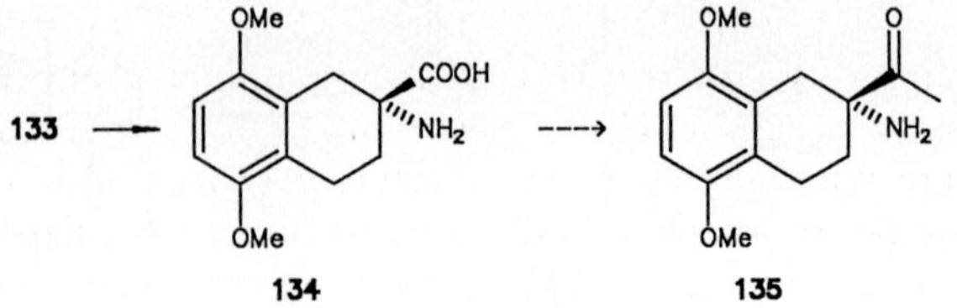

Scheme 37

The ethynyl cerium reagent has also been used to prepare the tetrahydro homophthalic anhydride building block (**139**). An interesting step is the cycloaddition of the allene (**136**) with 2-trimethylsiloxy-1,3-butadiene (**137**) to afford adduct (**138**). Chemoselective reaction with the dichloro cerium ethynyl reagent then affords the anhydride (**139**) after several steps using standard transformations (*132*) (Scheme 38).

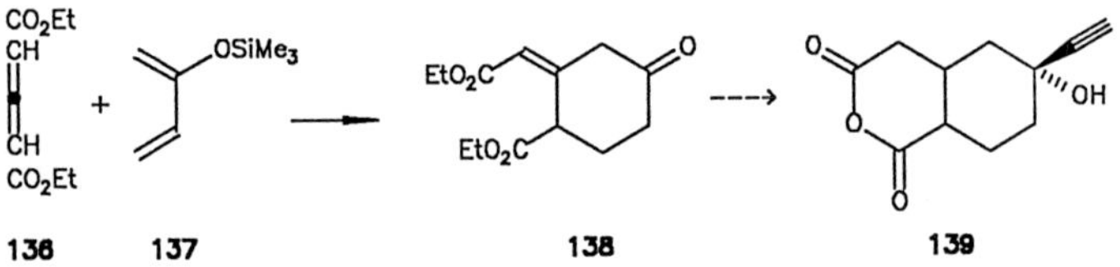

Scheme 38

The β-tetralone approach was also used in an asymmetric synthesis of quinone (**142**), which was then coupled with homophthalic anhydride to give (−)-7-deoxydaunomycinone (*139*). Grignard addition to the optically active acetal (**140**) – in which (−)-(2S,3S)-1,4-dimethoxy-2,3-butanediol is the chiral auxiliary – produced with essentially 100% de

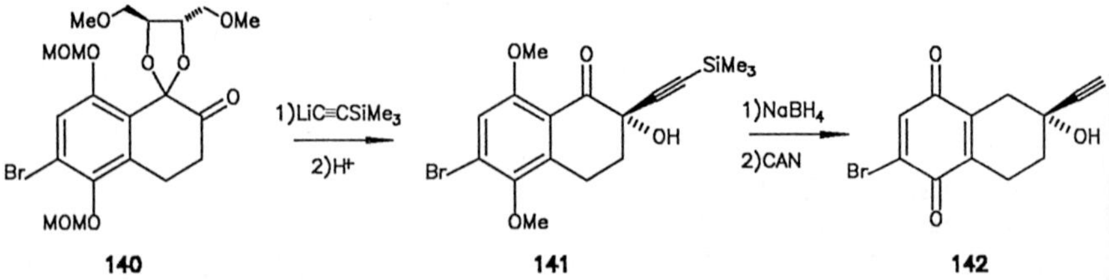

Scheme 39

the adduct (**141**). The benzylic carbonyl group in (**141**) can be reductively eliminated by borohydride reduction. Oxidation with ceric ammonium nitrate (CAN) then furnishes the optically active bromo quinone (**142**) (*139*) (Scheme 39).

Diels-Alder reactions of *ortho*-quinodimethanes (**144**) with methyl vinyl ketone (*10, 140, 141*) or more convergently with the silylenol ether (**145**) (*142*) can directly lead to the building block (**10**) (Scheme 40). The quinodimethanes can be obtained from bisbromomethyl benzenes such as (**143**) by treatment with zinc or sodium iodate but yields are especially high with chromium(II) chloride (*141*). An alternative ionic approach to (**10**) uses an intramolecular Grignard reaction of the bromo ketone (**146**) (*143*). The intramolecular Grignard (*144*) as well as the Diels-Alder reaction with *ortho*-quinodimethanes can also be applied successfully to tricyclic intermediates (*140, 145, 146*).

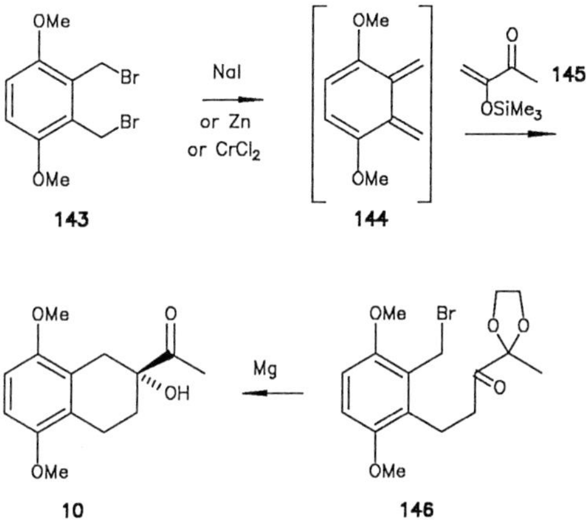

Scheme 40

The benzylic hydroxy group still missing in (**10**) can be introduced conveniently through reaction with methyl vinyl ketone of a benzofuran intermediate generated from precursor (**147**). The resulting adduct (**148**) (*147*) can also be obtained using carbanion chemistry (Scheme 41). The aldehyde (**149**) is obtained from the corresponding lithiated benzamide. Grignard reaction with (**150**) gives lactone (**151**) and DIBAH reduction followed by acetal cleavage and aldol condensation leads to the same intermediate (**148**) (*148*).

At this stage two effective methods for enantioselective synthesis of the important AB building block (**10**) starting from enone (**152**) (R = Me or Bn) deserve mention. The prochiral α, β-unsaturated ketone

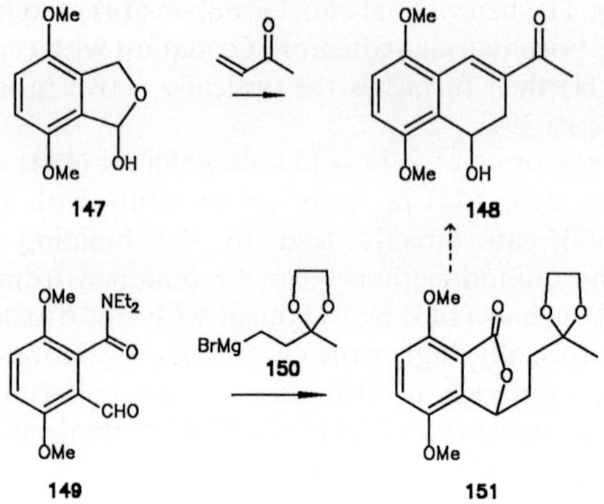

Scheme 41

(**152**) (R = Me) can be reduced enantioselectively with lithium aluminum hydride in the presence of (−)N-methyl ephedrine at −78 °C to give (**153**) (92% ee). Stereoselective SHARPLESS epoxidation yields (**154**) (90% de). The synthetic sequence is completed by LAH reduction of the epoxide (**154**) followed by FETIZONE oxidation (Ag_2CO_3) (*29*) (Scheme 42). Use of mixed methyl benzyl ethers [R = Me or Bn in (**152**)] allows generation of quinone monoacetals such as (**104**) or (**109**) for regioselective coupling with cyanophthalides (*149*). Interestingly, in substrates with an additional hydroxy group in (**153**) [e.g. the reduction product of (**148**)] the *homo*allylic hydroxy group controls the stereochemical outcome of the SHARPLESS epoxidation (*150*). A SHARPLESS reaction with racemic (**153**) was used for kinetic resolution (*112, 113, 149*), also Baker's yeast has been used for enantioselective reduction of (**152**) to (**153**) (*151*).

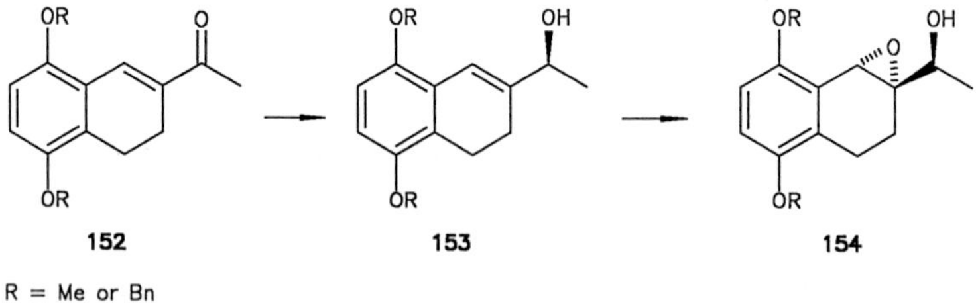

Scheme 42

Instead of the acid derived from (**97**), the unsaturated ketone (**152**) can be a substrate in a further extension of the asymmetric bromolactonization discussed earlier. This procedure avoids the additional step

References, pp. 77–88

of chain elongation. Acetalization of (**152**) with tartaric acid amide yields the chiral dioxolane (**155**) (*152*). Chirality transfer to the AB building block occurs upon bromolactonization with N-bromoacetamide to (**156**), which is further transformed in three steps to the enantiomerically pure A/B fragment (**10**) (Scheme 43).

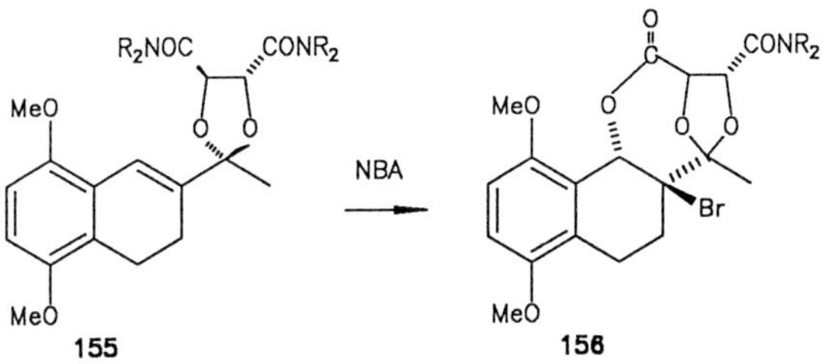

Scheme 43

Ketene acetals are particularly valuable partners for Diels-Alder reactions in anthracycline chemistry. Usually one alkoxy substituent is eliminated after the Diels-Alder addition. The remaining substituent introduces oxygen functionality into ring D or B respectively as shown in Scheme 10. A cyclic ketene acetal intended as a building block (**160**) for a convergent aklavinone synthesis (*153*) was prepared by RAPOPORT et al. (*154*) (Scheme 44). The key step in its preparation was the one-step Michael addition-elimination with subsequent intramolecular Darzens glycidic ester condensation between dimethyl (3-oxopentyl)malonate (**157**) and methyl α,β-dichlorocrotonate (**158**) which led to epoxide (**159**). This was further transformed into diene (**160**) by demethoxycarbonylation and ketene acetyl formation on base treatment and quenching with trimethylchlorosilane (*155*).

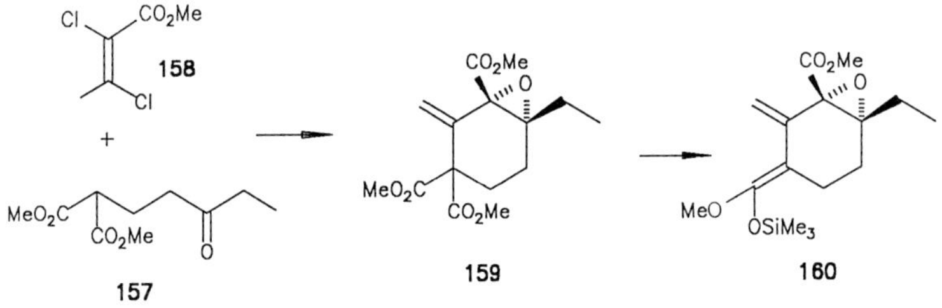

Scheme 44

Asymmetric synthesis of an AB ring fragment (**163**) of aklavinone was realized by MEYERS and HIGASHIYAMA (*156*) (Scheme 45). Although the fragment cannot be directly incorporated into anthracyclinones a potential application on a tetracyclic level could be possible. Chirality transfer occurs during the addition of vinyllithium to the chiral oxazolidine (**161**) with 87% de. The nucleophilic aromatic addition is also chemically interesting because the normal regenerative behaviour of the aromatic system does not occur and the dihydronaphthalene (**162**) is isolated. The AB fragment (**163**) is then obtained in ten additional steps. The absolute configuration of this fragment was opposite to the natural configuration of aklavinone.

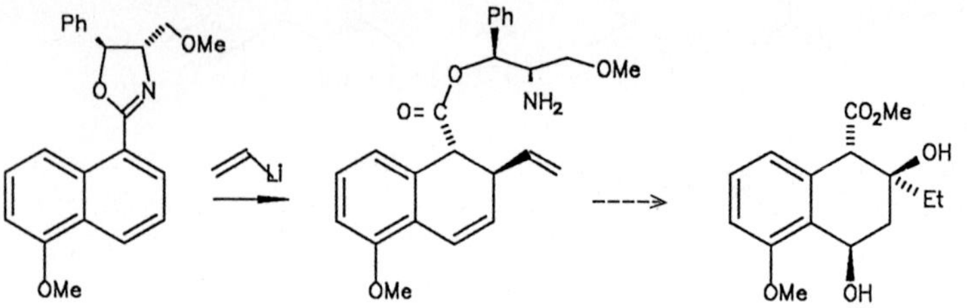

Scheme 45

Building blocks with an exocyclic ester group such as (**159**) may be used not only as precursors of ketene acetals for Diels-Alder reactions but also for ionic couplings similar to that shown in Scheme 16. Starting from commercially available tetrahydrophthalic ester (**164**) PENCO et al. (*157*) were able to synthesize the AB building block (**167**), a precursor for the synthesis of 6-deoxydaunomycinones (Scheme 46). Interestingly the organometallic reagent used in the coupling reaction adds to the ester and not to the lactone group of (**167**). In the synthesis of (**167**) acylation to (**165**) is followed by treatment with tosylhydrazone

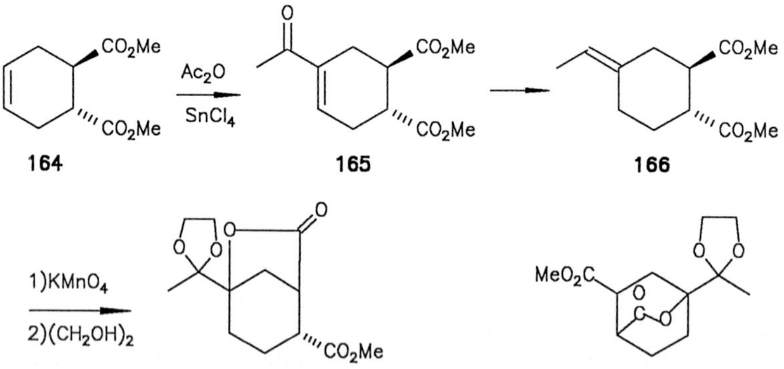

Scheme 46

and elimination to the olefin (**166**). The *cis*-hydroxylation of (**166**) with potassium permanganate is followed by spontaneous lactonization and ketalization to (**167**) (*157*). A similar reaction sequence leads to the regioisomer (**168**) which is a building block for 11-deoxydaunomycinones (*158, 159*).

IV. Construction of Ring A

1. Diels-Alder Reactions

Annelation of the C_4-fragment with anthraquinone derivatives to produce ring A can be carried out by Diels-Alder as well as electrophilic or nucleophilic reactions. The Diels-Alter reaction of anthradiquinone (**169**) (Scheme 47) is complicated by the possibility of addition to the internal or external double bond (*160, 161, 162*). However, the formation of linearly condensed systems (**171**) predominates with 2-acetoxybutadiene (**170a**) (*163*) or chloroprene (**170b**) (*164*) [for related reactions compare (*37, 162*)]. Tetracyclic ketones such as (**171**), including 4-, 6-, and 11-deoxy analogues are further transformed into anthracyclines by addition of organometallic reagents in a similar manner to that shown for their bicyclic counterparts in Scheme 36 (*36, 70, 78, 80, 107, 134, 135, 137, 162, 163, 165, 166, 167, 168, 169, 170, 171, 172, 173, 174*).

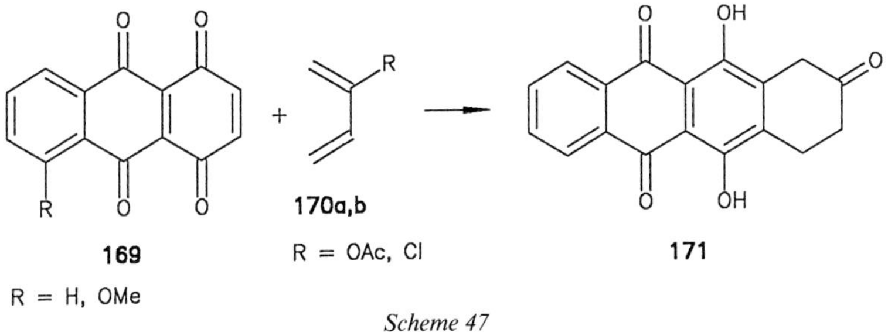

169
R = H, OMe

170a,b
R = OAc, Cl

171

Scheme 47

Several solutions to the problem of internal vs. external addition to the anthradiquinone double bond have been proposed. 1,4-Anthraquinones give linearly condensed products exclusively; however additional oxidation steps are required for their conversion to anthracyclinones (*141, 175, 176, 177, 178*). A more successful strategy involves successive Diels-Alder additions to naphthazarine as shown in

Scheme 8 (*35, 36, 37, 179, 180, 181, 182*). If 1,3-bis(trimethylsiloxy)-1,3-butadiene (**31**) is used as diene simultaneous introduction of two oxygen functions into ring A and further transformation to fully functionalized daunomycinones is possible (compare Scheme 8) (*35, 180, 181*). The strategy of two subsequent Diels-Alder additions has also been used by VOGEL *et al.*, the central unit being the tetramethylidene-oxabicyclo-[2.2.1]heptane (**172**) (Scheme 48). Control of regiochemistry is also possible to some extent using this building block. Scheme 48 illustrates the successive construction of the tetracyclic skeleton starting from (**172**) to afford (**173**) and in the next addition step (**174**) (*183*).

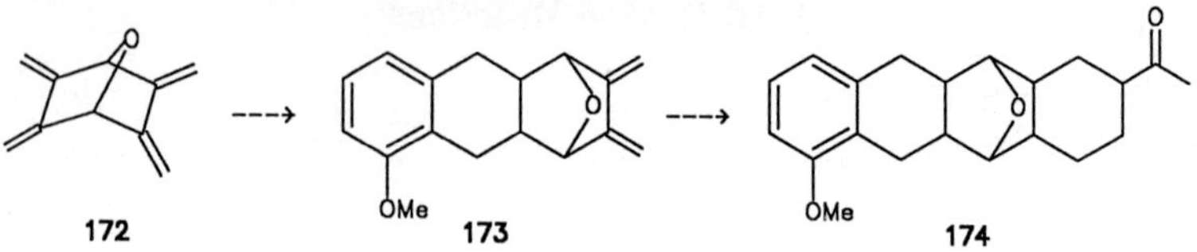

Scheme 48

STOODLEY and coworkers have introduced a third strategy for synthesis of linearly condensed tetracycles starting from anthradiquinone epoxides such as (**175**) (*14, 37, 184, 185*) (Scheme 49). Of particular value is the enantioselective synthesis of (+)-4-demethoxydaunomycinone in which the glucose coupled diene (**176**) is reacted with (**175**). The carbonyl group of ring A in (**177**) is generated upon acidic cleavage of an intermediate silylenol ether and is transformed in a similar fashion

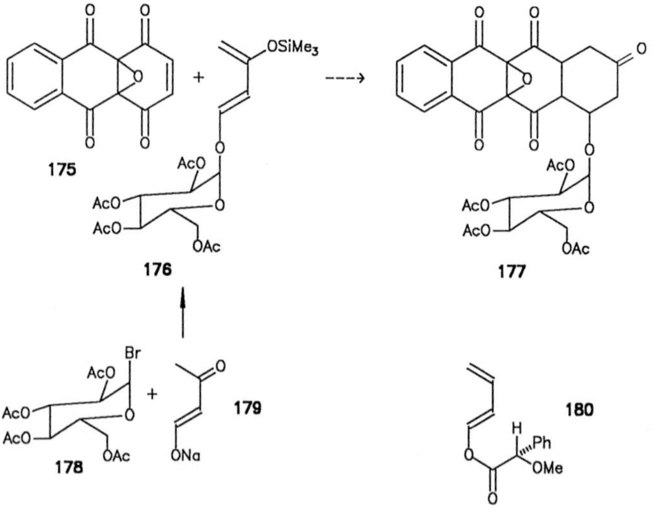

Scheme 49

References, pp. 77–88

into the tertiary alcohol of daunomycinone as shown for bicyclic systems in Scheme 36. The diastereoselection [4:1 in favor of (**177**)] observed in the cycloaddition of the epoxytetrone (**175**) and the chiral diene (**176**) is remarkable (*184*). There are very few examples in which chiral dienes show satisfactory diastereoselection, the most notable example being diene (**180**) of TROST (*186*). The diene (**176**) is very easily prepared, although in low yield (ca. 20%), from the readily available starting materials acetobromoglucose (**178**) and sodium salt (**179**).

2. Electrophilic Additions

The synthesis of rhodomycinones with *alkyl* side chains using the Marschalk reaction has already been shown in Scheme 11. The electrophilic addition of aldehydes to anthrahydroquinones can also be looked upon as a kind of aldol reaction and takes place at about 0 °C with 1,4-**di**hydroxyanthrahydroquinones. However, only nonenolizible aldehydes add to **mono**hydroxyanthrahydroquinones in *inter*molecular reactions (*187*). Evidently both hydroxy groups at C-1 and C-4 are needed for the annelation of ring A if alkyl groups for the attachment of side chains are missing. Aloe-emodin is a good starting material for synthesis of 11-deoxyanthracyclines because its hydroxymethyl group allows the construction of side chains that can undergo *intra*molecular Marschalk reaction to 11-deoxydaunomycinone (*188*).

Twofold Marschalk reactions with 1,4-dihydroxyanthrahydroquinones are ideal for adaptation to the synthesis of enantiomerically pure rhodomycinones. To this end it is necessary to prepare chiral building blocks as masked 1,4-dialdehydes. The aldehydes (**71**) and (**75**) shown in Scheme 21 (*87*) can be used not only for construction of AB building blocks but were also directly coupled to leucoquinizarin (*189*). However, only derivatives with a hydroxymethyl side chain can be obtained from these building blocks. More recently isosaccharino-1,4-lactone (**70**), has been used as starting material for chiral building blocks with an ethyl side chain (*53*) (Scheme 50). Transformation of the hydroxymethyl group involved mesylation of (**181a**) to (**181b**) and reaction with lithium dimethyl cuprate in ether followed by acetalization to (**182**). Selective acetal and periodate cleavage yields the aldehyde (**183**). Coupling to leucoquinizarin, acetal cleavage to (**184**), oxidation of the primary hydroxy group to an aldehyde by the MOFFAT procedure and cyclization completes the synthesis to 4-deoxy-γ-rhodomycinone (**185**).

Glucose can be the starting material for another chiral building block (**188**). The ethyl group is introduced in a stereospecific reaction

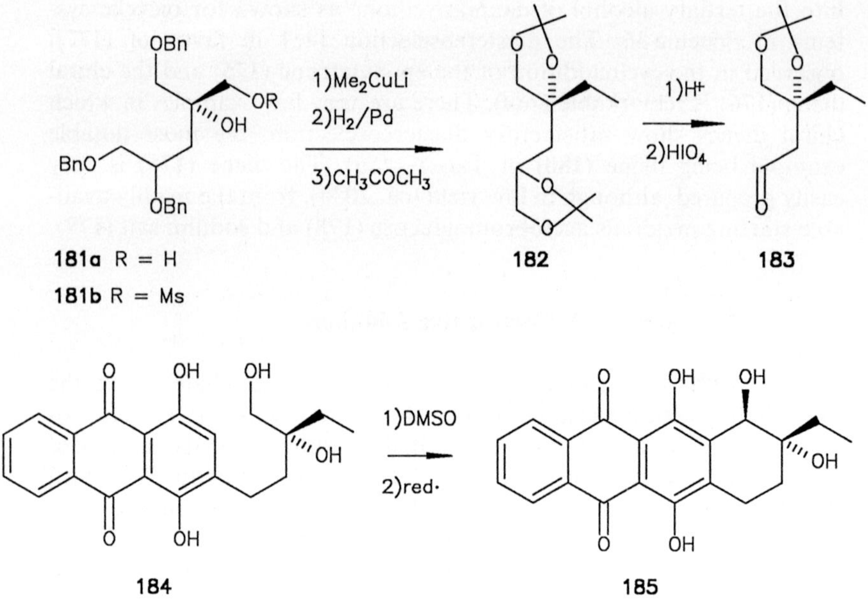

Scheme 50

181a R = H
181b R = Ms

Scheme 50

of keto sugar (**186**) with ethylmagnesium bromide (*54*). The adduct is first benzylated to (**187**). Acetyl and glycoside cleavage affords a hydroxy aldehyde that is represented in the cyclic form (**188**) (Scheme 51).

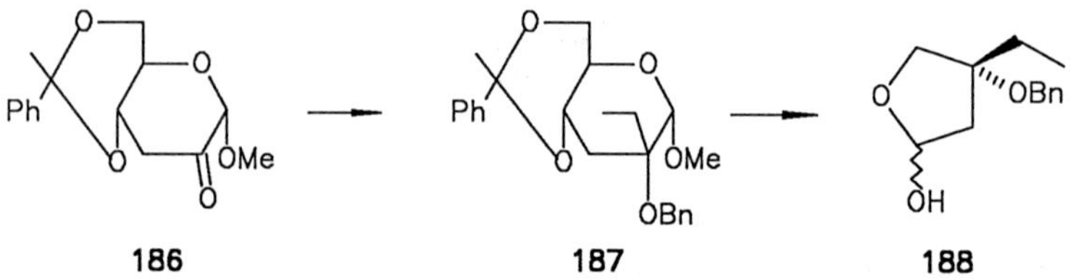

Scheme 51

Simple monosaccharides cannot be coupled to anthrahydroquinones under the normal conditions of the Marschalk reaction (*190*). SHAW et al. (*52*) have used selective cleavage of diacetone glucose to synthesize aldehydes that can be coupled to leucoquinizarin to afford anthracyclinone analogues without side chain (Scheme 52). The ethynyl group can be introduced by reaction of the well known keto sugar (**189**) with ethynyl magnesium bromide affording (**190**) after benzylation. Marschalk reaction with an aldehyde derived from (**190**) followed by glycol cleavage and a second Marschalk reaction yields the all-*cis* tetraol (**191**). A major difference between these products and normal anthracyclinones is the presence of the additional hydroxy group at C-8.

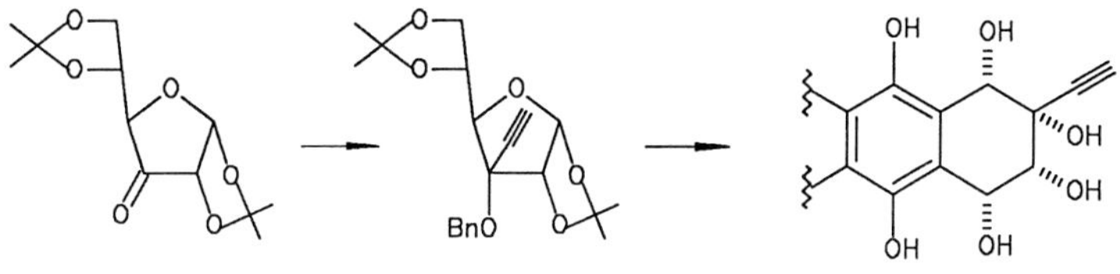

Scheme 52

Chiral building blocks derived from (S)-lactic acid (**192**) (*55*) and the commercially available (S)-2-aminobutyric acid (**195**) (*56*) can be prepared in a strategy of self-reproduction of chirality similar to the one used in the synthesis of the AB building block (**85**) (Scheme 22). (S)-lactic acid is transformed to the pseudolactone (**193**) by acetalization and subsequent allylation with allyl bromide. DIBAH reduction of (**193**) gives an α-hydroxy aldehyde (**194**) that can be coupled to leucoquinizarin in a similar way as described for (**183**) (*55*) (Scheme 53).

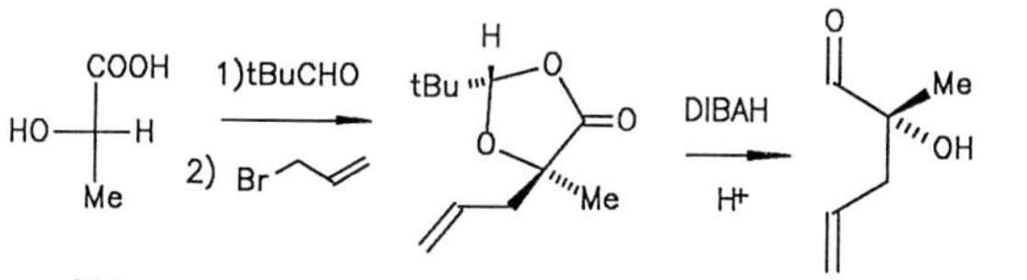

Scheme 53

However, lactic acid leads only to pharmaceutically less interesting anthracyclinones with a methyl instead of an ethyl side chain. For synthesis of the ethyl analogues, optically active α-hydroxy butyric acid was synthesized from the commercially available (S)-2-amino butyric acid by treatment with nitrous acid. Acetalization and allylation then gave (**196**). A decisive progress in convergency was made with transacetylation of the DIBAH reduction product of (**196**) to afford (**197**). In the crystalline ozonolysis product (**198**) all the requisite functionalities are present and the regioselective coupling with 1,4,5-trihydroxy-anthraquinone gave the advanced precursor (**199**) [7:1 in favour

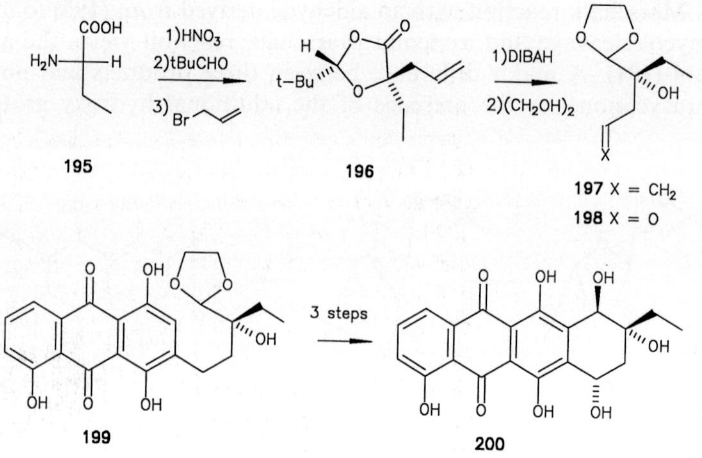

Scheme 54

of (**199**)]. The first total synthesis of enantiomerically pure β-rhodomycinone (**200**) was completed in only three further steps (56) (Scheme 54).

3. Nucleophilic Additions

Nucleophilic additions to 9,10-anthraquinones can be regarded as phenylogous Michael additions to the quinone carbonyls. The equilibrium represented by (**201a**) and (**201b**) is possible with 1,4-dihydroxy-9,10-anthraquinones. Tautomer (**201b**) may be present in very low con-

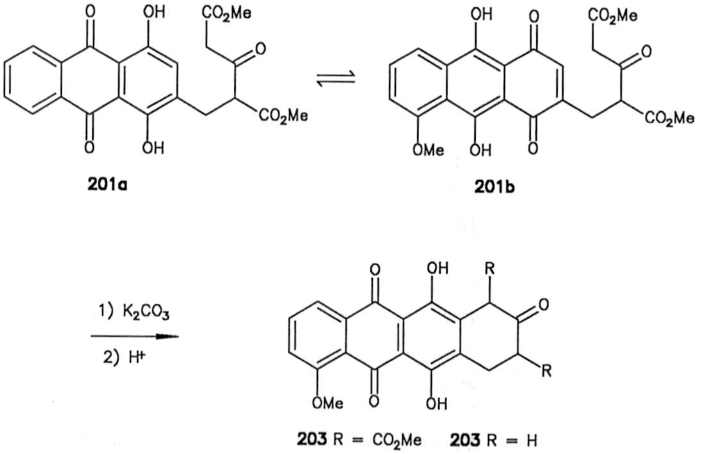

Scheme 55

centrations that are not detectable by NMR spectroscopy; nucleophilic additions are nevertheless enormously accelerated. One example is the very easy cyclization of (**201b**) to the tetracycle (**202**) by potassium carbonate in methanol. Decarboxylation under acidic conditions provides a very short and regioselective synthesis of the important (*163*) daunomycinone precursor (**203**) (*173*) (Scheme 55). Precursors similar to (**201**) but without the hydroxy group at C-4 cyclize only under drastic reaction conditions (NaH, DMF, 80 °C) and the reaction products are always the fully dehydrated naphthacenequinones (*191*).

An equilibrium similar to that shown in Scheme 55 must also play an important role in the carminomycinone syntheses of SUTHERLAND et al. (*59*) (Scheme 12). Similarly a nitromethane fragment can be inserted into the readily available ketones (**204**) in a one step anthracyclinone synthesis (*192*) (Scheme 56). The nitromethane presumably adds to the carbonyl group and subsequently to the anthraquinone moiety after which the nitro group is eliminated in an intramolecular redox reaction to afford (**205**). Interestingly, the hydroxy group at C-1 is not required for the cyclization.

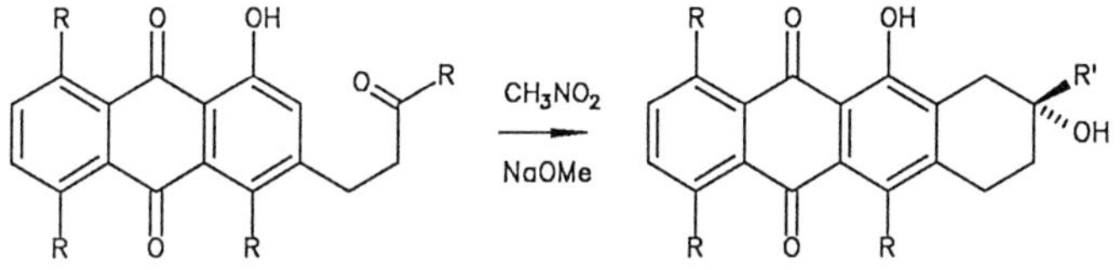

204   R = H or OH, R' = Me or Et   **205**

Scheme 56

A phenolic group *ortho* to the ring closure position is not always necessary for cyclization as CAVA et al. (*193*) were able to show (Scheme 57). When a cyanosulfide was used as a nucleophile in a tandem Michael-aromatic substitution reaction with the α,β-unsaturated ketone (**206**). Elimination again took place (in this case of thiophenolate) after cyclization and the aklavinone analogue (**207**) was isolated in 83% yield. The intramolecular nucleophilic addition is in this particular case certainly facilitated by the additional electron withdrawing effect of the benzylic carbonyl group. However, the tertiary hydroxy group of aklavinone may be difficult to introduce into (**207**) and the ketoester cyclization shown in Scheme 15 offers advantages for ε-rhodomycinone (**66**), aklavinone (*40, 62, 63, 194*), pyrromycinone (*63, 195*), auramycinone (*195, 196*) and nogalamycinone (*65*) synthesis.

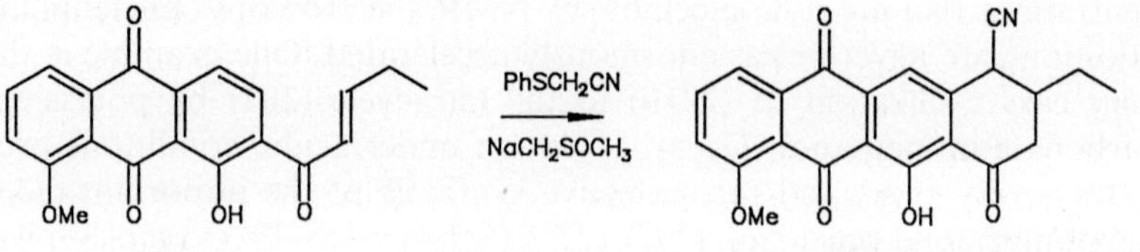

Scheme 57

V. Building Blocks for Rings C and D

The aromatic D or CD fragments of the anthracyclinones are much easier to construct than the chiral AB building blocks and only a few points pertaining to this aspect of anthracyclinone synthesis will be mentioned in this chapter.

Dienes are used for annelation of ring D with quinoide tricyclic intermediates such as (210) (only the reactive tautomer is shown in Scheme 58). Alkoxy groups at the C-1 position of the open chain diene 1-methoxy-1,3-butadiene (209) are easily eliminated from the primary Diels-Alder adducts thermally or by treatment with base. In this case the 4-demethoxydaunomycinone (211) is obtained in high yield (*180*). With the cyclic diene 1-methoxy-1,3-cyclohexadiene (208) the methoxy substituent can be preserved in a Diels-Alder/oxidation/*retro* Diels-Alder reaction sequence to afford daunomycinone (212) (*35, 180, 181*).

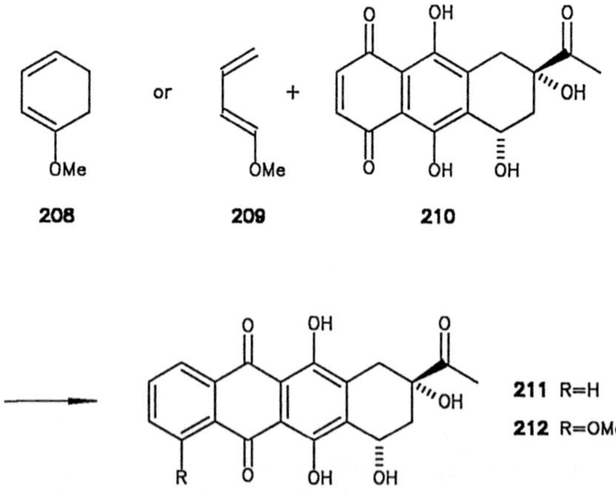

211 R=H
212 R=OMe

Scheme 58

Very often simple metalated aromatic systems are coupled with AB fragments containing aldehyde or ester groups (*107*) (see Scheme 14). The metalated species can be obtained by halogen-metal exchange or by *ortho*-directed metalation (*68, 147,* review *197*).

Reaction partners in the Friedel-Crafts reactions shown in Scheme 2 are phthalic acid derivatives. In some cases the requisite hydroxyphthalide can be obtained by reduction of the corresponding phthalic acid anhydrides with lithium tri-*tert*-butoxy aluminum hydride (*198*). Normally the sterically less hindered carbonyl group is reduced preferentially in substituted systems. The hydroxyphthalides are precursors of the cyanophthalides (*120, 174, 199, 200*) and also sulfonylphthalides (*63, 64, 71, 122, 201, 202, 203*), which permit excellent control of regiochemistry in the construction of the anthracyclinone skeleton (see Scheme 15).

In addition, flash vacuum pyrolysis of hydroxy phthalides such as (**213**) (Scheme 59) yields the benzocyclobutanediones (**214**) which can be coupled photochemically in high yield with the enantiomerically pure building block (**85**) [(*88*) Scheme 22] to afford the optically active 1-methoxy-daunomycinone (**215**) in only two steps (*92*). The mild conditions of the photochemically-induced coupling reaction permits the use of fully functionalized optically active building blocks. A limitation is the lack of regiocontrol with unsymmetrically substituted benzocyclobutanediones (*204*).

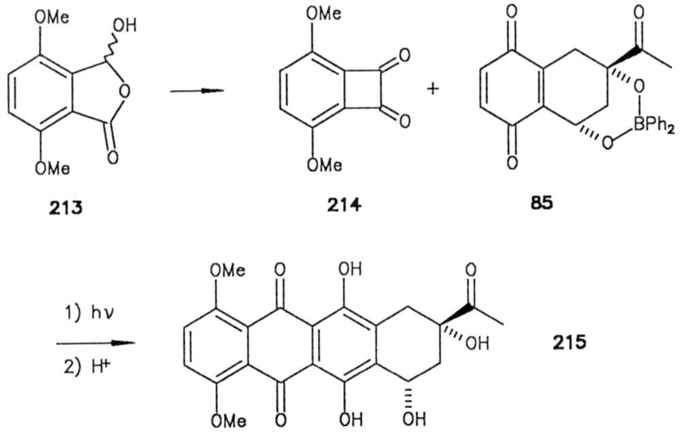

Scheme 59

The importance of the substituted phthalic acid derivatives justifies a brief comment on their synthesis. A general synthetic method is the introduction of a formyl group in the aromatic system by directed metalation of appropriate benzamide precursors (review see *197*). Thus, the benzamide (**216**) (*109, 148*) can be metalated by treatment with *sec*-butyl lithium. Quenching of the lithium compound with DMF and subsequent acidic hydrolysis then affords the corresponding hydroxyphthalide (**217**). Generation of the cyanohydrins with hydrogen cyanide presents no difficulty but the cyclization to the cyanophthalides (**218**) sometimes gives low yields. The Vilsmeyer complex generated from oxalyl chloride and DMF was found to be a good reagent for the cyclization step of the cyanohydrins to the cyanophthalides (*109*). Dicyclohexylcarbodiimide (DCC) has also been successfully applied in the cyclization step (*200*) (Scheme 60). Very high yields are obtained by reaction of the formyl-N,N-diethylbenzamides with trimethylsilyl cyanide and acetic acid (*205*). The analogous sulfonylphthalides can be obtained by sulfenylation of the corresponding phthalides in contrast to the cyanophthalides (*206*). The hydroxyphthalides are also available by bromination and hydrolysis of the appropriate phthalides (*200*).

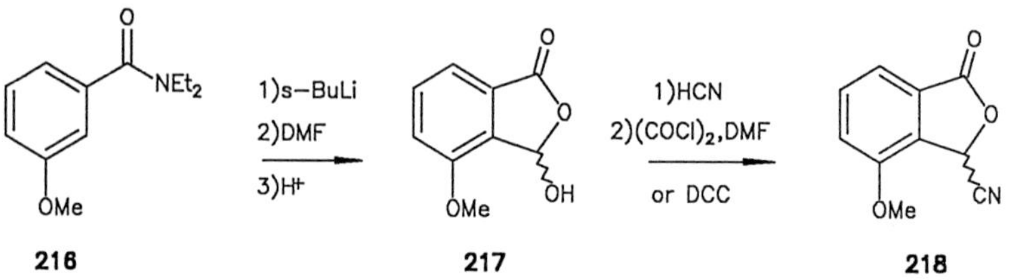

Scheme 60

Related trimethoxy derivatives (*207*) and as well as fluorinated DC building blocks (*109, 200*) are readily available by the directed metalation strategy. However, if the aromatic system possesses competitive metalation sites alternative methods have to be employed. One way of introducing the one-carbon unit is the Michael addition of cyanide to quinone esters such as (**219**). The ester group directs the regioselective introduction of cyanide. The trimethyl ether (**220**) is isolated after methylation of the intermediate hydroquinone from the cyanide addition (*207*) (Scheme 61).

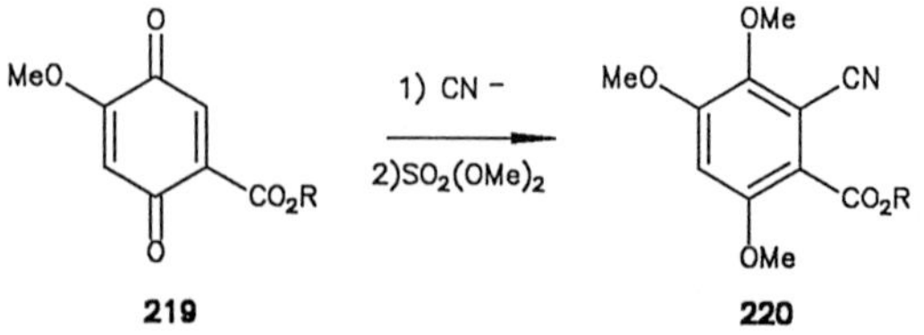

Scheme 61

References, pp. 77–88

Alternatively, the addition of oxygen functions to simple aromatic esters can be used to construct more complex substitution patterns. However, only modest yields were obtained in the Elbs oxidation of appropriate precursors (e.g. ethyl evernate) (*207*). A better method is the Thiele-Winter reaction of the benzoquinone diester (**221**). The trimethoxy phthalic ester (**222**) can be isolated in 56% overall yield by treatment of quinone (**221**) with acetic anhydride, methanolysis and methylation (*207*) (Scheme 62).

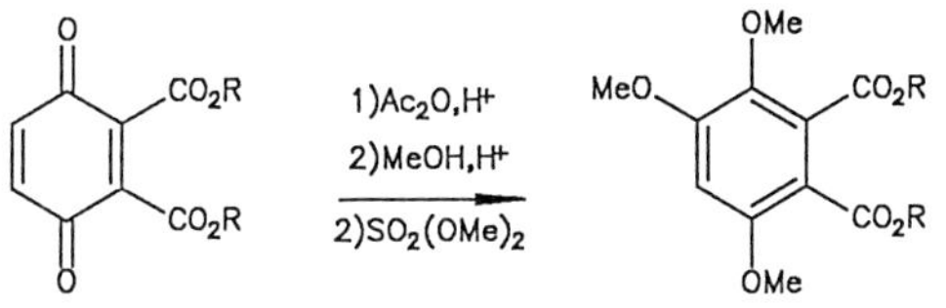

Scheme 62

Symmetrically substituted phthalic esters or hydroxyphthalides can be synthesized by Diels-Alder reaction of substituted cyclohexadienes with acetylene carboxylic esters. The 1,3-cyclohexadiene (**223**) is available by Birch reduction (*208*) and *in situ* isomerization with *tris*(triphenylphosphine)chlororhodium (*209*). Diels-Alder reaction with the acetylenic acetal ester (**224**) (*210*) initially affords the bicyclo[2.2.2]octadiene (**225**), which then decomposes to ethylene and aromatic compounds (*109*) (Scheme 63). Acidic hydrolysis gives the desired hydroxyphthalide (**213**) and subsequent flash vacuum pyrolysis the benzocyclobutadione (**214**). The corresponding phthalic ester is obtained from (**223**) and dicarboxylic ester (*209*). Birch reduction and isomerization of 1,2,4-trimethoxybenzene gives too many isomers and this limits the Diels-Alder strategy to symmetrically substituted phthalic ester derivatives (*207*).

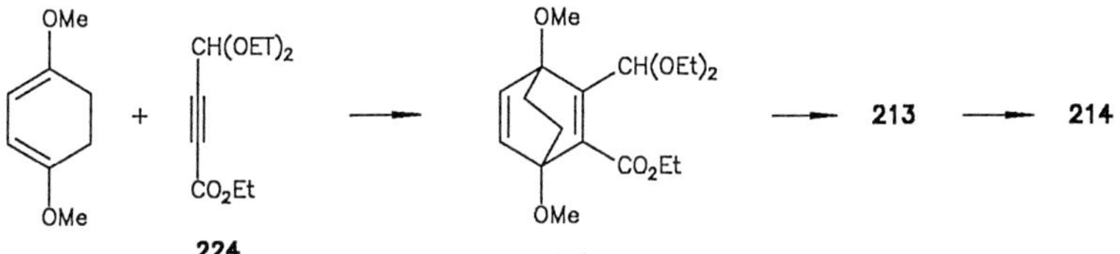

Scheme 63

Not only derivatives of phthalic esters but also those of *homo*phthalic anhydrides may serve as CD building blocks for anthracyclinone synthesis. Of special use are those homophthalic anhydrides which are methoxylated or acetoxylated at the benzylic position. Thus, an additional oxidation step in the conversion of a late intermediate to 9,10-anthraquinones is avoided. The CD building block **(227)** is best obtained by acetoxylation at the benzylic position of the bis(trimethylsilyl)ketene acetyl derived from the homophthalic acid **(226)** followed by anhydride formation *(210a)*. The strong base (NaH) induced cycloaddition is exemplified by reaction of **(227)** with the chloroquinone derivative **(228)** to yield regioselectively the adduct **(229)** *(210a)*. Treatment of **(229)** with trifluoroacetic acid converts the 1,4-anthraquinone to the known daunomycinone intermediate **(203)** (Kende-ketone) *(163)*. The mechanism has not yet been elucidated and the reaction may proceed either by Diels-Alder or by ionic cycloaddition.

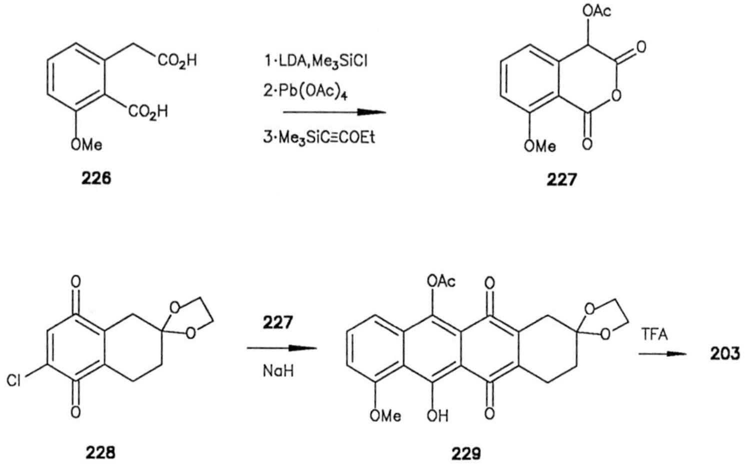

Scheme 64

VI. Concluding Remarks

Today almost any naturally occurring anthracyclinone and many purely synthetic derivatives including those deoxygenated in the anthraquinone moiety *(125, 126, 211, 212)*, heterosubstituted derivatives *(179, 213)* or hetero-analogues *(83, 95, 214, 215, 216)* can be obtained by total synthesis. Moreover the international competition devoted to the chemistry of this clinically important class of compounds has contributed to the general improvement of synthetic methods. Notably many new solutions have been found to the problems involved in construction of linearly condensed ring systems. In particular regioselective introduction of remote substituents and synthesis of enantiomerically pure building blocks either by asymmetric induction or by the incorporation of fragments derived from the "chiral pool" have found many new applications. Enantioselectivity of anthracyclinones of type B with an ester group at C-10 is still a certain problem. An enantioselective variation of the chemically effective ketoester cyclization (*66*) that has found general application for this particular type of anthracyclinones (*13*) has been realized only once in the aklavinone synthesis of KISHI (*194*).

Industrial laboratories are, however, reluctant to apply the more advanced techniques of academic research. Thus, resolution of racemic mixtures (*16, 102*) is preferred to asymmetric synthesis, traditional Friedel-Crafts reactions which result in longer overall reaction sequences are often employed (*16, 20*), and most symmetrically substituted products are synthesized by means of Diels-Alder reactions (*102*). A major drawback to adaption of the more elaborate syntheses for larger scale production of the tetracyclic products, which are difficulties of manipulation due to extremely poor solubility in organic solvents, especially in the β-rhodomycinone series. In this area highly selective syntheses which avoid the need for chromatographic separations are still required.

With few exceptions (*138*) anthracyclinones are only active as glycosides with a number of 2-deoxysugars attached to OH-7 and or to OH-10 or C-glycosidically to ring D. Many semisynthetic derivatives are prepared by modification of the sugar part of the molecule (*7, 10*) and progress has been reported in the synthesis of the C-glycoside in the noglamycine series (*217*). Newly prepared aglycones are usually tested as easily obtainable monoglycosides. However, many derivatives show improved activity as di- or trisaccharides especially in the rhodomycinone series. Although much work has been done on synthesis of the di- and trisaccharide fragments (*10*), effective methods for coupling of the oligosaccharides to the aglycones are still wanted and present an enormous challenge for future investigations.

References

1. BROCKMANN, H.: Anthracycline und Anthracyclinone. Fortsch. Chem. Org. Naturst. **21**, 121 Wien: Springer-Verlag. 1963.
2. THOMSON, R.H.: Naturally Occurring Quinones III. London, New York: Chapman and Hall 1987.

3. HARRINGTON, A.A., and A.D. ROBERTS: Dictionary of Antibiotics and Related Substances. Ed. B.W. BYCROFT, p. 30. London, New York: Chapman and Hall. 1987.
4. ARCAMONE, F.: Doxorubicin. New York: Academic Press. 1980.
5. GESSON, J.-P., and M. MONDON: Synthèse Totale et Activité Cytotoxique de Nouvelles Anthracyclines. Actual. Chim. Thér. **13. serie,** 161.
6. BROWN, J.R., and S.H. IMAM: Recent Studies on Doxorubicin and its Analogues. In: Prog. Med. Chem. Eds. G.P. ELLIS and G.B. WEST, Vol. 21, pp. 170–236, 1984.
7. Anthracyclines. Current Status and New Developments. Eds. S.T. CROOKE and S.D. REICH. New York: Academic Press. 1980.
8. ARCAMONE, F.: Properties of Antitumor Anthracyclines and New Developments in Their Application: Cain Memorial Award Lecture. Cancer Res. **45**, 5995 (1985).
9. BRAZHNIKOVA, M.G., V.B. ZBASKIY, N.P. POTANOVA, Y.N. SCHEINKER, T.F. VLASOVA, and B.V. ROZYNOV: Antibiotiki (Moskow) **18**, 1059 (1973).
10. EL KHADEM, H.S.: Anthracycline Antibiotics. New York, London: Academic Press. 1982
11. KELLY, T.R.: Synthetic Approaches to Anthracycline Antibiotics. Annu. Rep. Med. Chem. **14**, 288 (1979).
12. BROADHURST, M.J., C.H. HASSALL, and G.J. THOMAS: Synthesis of Anthracyclinones. Chem. and Ind. **1985**, 106.
13. KROHN, K.: Total Synthesis of Anthracyclinones. Angew. Chem. **98**, 788 (1986); Angew. Chem. Int. Ed. **25**, 790 (1986).
14. STOODLEY, R.J.: The Anthracycline Antibiotics: Some Synthetic Endeavours. Second SCI/RSC Medical Chemistry Symposium; Spec. Publ.-R. Chem. Soc. **1984**, 134.
15. BAYER, O.: Anthrachinone. Ed. HOUBEN-WEYL-MÜLLER, Vol. VII/3c. Stuttgart: Thieme. 1979.
16. ARCAMONE, F., L. BERNARDI, B. PATELLI, P. GIARDINO, A. DIMARCO, A.M. CASAZZA, C. SORANZO, and G. PRATESI: Synthesis and Antitumor Activity of new Daunorubicin and Adrinamycin Analogues. Experientia **34**, 1255 (1978).
17. WONG, C.M., D. POPIEN, R. SCHWENK, and J. TE RAA: Synthetic Studies of Hydronaphthacenic Antibiotics. I. The Synthesis of 4-Demethoxy-7-O-methyl Daunomycinone. Can. J. Chem. **49**, 2712 (1971).
18. KROHN, K., and B. BEHNKE: Synthetische Anthracyclinone, XV. Regio- und stereoselektive Synthese der α-, β- und γ-Rhodomycinone über intramolekulare Marschalk-Cyclisierung. Chem. Ber. **113**, 2994 (1980).
19. MATSUMOTO, T., M. OHSAKI, M. SUZUKI, Y. KIMURA, and S. TERASHIMA: Synthesis of 4-Demethoxyanthracyclines Carrying a Lipophilic Alkanoyl Group at the C_9-Position. Chem. Pharm. Bull. **34**, 4605 (1986).
20. CONFALONE, P.N., and G. PIZZOLATO: Total stereospecific Synthesis of ($\pm$)-Aklavinone. J. Am. Chem. Soc. **103**, 4251 (1981).
21. GLEIM, R.D., S. TRENBEATH, F. SUZUKI, and C.J. SIH: Regiospecific Syntheses of Islandicin and Digitopurpone Monomethyl Ethers. J. Chem. Soc., Chem. Commun. **1978**, 242.
22. BROADHURST, M.J., and C.H. HASSALL: Anthracyclines. Part 1. The Synthesis of Racemic Daunomycinone and Some related Tetrahydronaphthacenequinones. J. Chem. Soc. Perkin Trans. 1 **1982**, 2227.
23. BRAUN, M.: Regioselektive Synthese von Daunomycinon und γ-Rhodomycinon. Tetrahedron Lett. **21**, 3871 (1980).
24. – A Regioselective Synthesis of Daunomycinone and Related Anthracyclinones. Tetrahedron **40**, 4585 (1984).
25. BRAUN, M., R. VEITH, and G. MOLL: Regioselektive Addition von Grignard-Reagentien an 3-Methoxy- und 3-Nitrophthalsäureanhydride. Chem. Ber. **118**, 1058 (1985).
26. PARKER, K.A., and J. KALLMERTEN: Efficient, Regiospecific Synthesis of Anthracyc-

line Intermediates: Total Synthesis of Daunomycinone. J. Am. Chem. Soc. **102**, 5881 (1980).
27. — Approaches to Anthracyclines. 1. Conjugate Aroylation of α,β-Unsaturated Esters. J. Org. Chem. **45**, 2614 (1980).
28. — Approaches to Anthracyclines. 2. Regiospecific Annelative Quinone Synthesis. J. Org. Chem. **45**, 2620 (1980).
29. TANNO, N., and S. TERASHIMA: Asymmetric Synthesis of Optically Active Anthracyclinone Intermediates and 4-Demethoxyanthracyclinones by the Use of a Novel Chiral Reducing Agent. Chem. Pharm. Bull. **31**, 821 (1983).
30. RUSSELL, R.A., R.W. IRVINE, and A.S. KRAUSS: A Caveat Regarding Chiroptical Measurements of Chiral Anthracyclinones. Tetrahedron Lett. **25**, 5817 (1984).
31. WELZEL, P.: Asymmetrische Diels Alder Reaktionen. Nachr. Chem. Tech. Lab. **31**, 979 (1983).
32. KROHN, K.: Asymmetrische Induktion bei Diels-Alder Reaktionen. Nachr. Chem. Tech. Lab. **35**, 836 (1987)
33. TAMIREZ, J., and P. VOGEL: Macrocycles by Intramolecular Diels-Alder Reaction – Regioselective Synthesis of Anthracycline Precursors. Angew. Chem. **96**, 61 (1984); Angew. Chem. Int. Ed. **23**, 74 (1984).
34. BRESLOW, R., R.J. CORCORAN, B.B. SNIDER, R.J. COLL, P.L. KHANNA, and R. KALEYA: Selective Halogenation of Steroids Using Attached Aryl Iodide Templates. J. Am. Chem. Soc. **99**, 905 (1977).
35. KELLY, R.T., L. ANANTHASUBRAMANIAN, K. BORAH, J.W. GILLARD, R.N. GOERNER JR., P.F. KING, J.M. LYDING, W.-G. TSANG, and J. VAYA: An Efficient Regioselective Synthesis of ($\pm$)-Daunomycinone. Tetrahedron **40**, 4569 (1984).
36. ECHAVARREN, A., P. PRADOS, and F. FARINA: Polycyclic Hydroxyquinones-XIX. Regiospecific Synthesis of Anthracyclinones via the Diels-Alder Reaction with Dichloronaphthazarins. Tetrahedron **40**, 4561 (1984).
37. POTMAN, R.P., N.J.M.L. JANSSEN, J.W. SCHEEREN, and R.J.F. NIVARD: Application of 1-tert.-Butoxy-3-[(trimethylsil)oxy]buta-1,3-diene in the Preparation of Functionalized β-Hydroxycyclohexanone Derivatives, Including Valuable Precursors of Daunomycinone Analogues. J. Org. Chem. **52**, 3628 (1984).
38. KROHN, K., and K. TOLKIEHN: Stereoselektive Synthese des 4-Desmethoxydaunomycinons. Tetrahedron Lett. **1978**, 4023.
39. KROHN, K.: Synthetische Anthracyclinone, XVI. Synthese hydroxylierter Anthrachinone durch regioselektive Diels-Alder Reaktion. Tetrahedron Lett. **21**, 3557 (1980).
40. – Synthetische Anthracyclinone, XXIII. Synthese und Konfiguration der stereoisomeren Aklavinone. Liebigs Ann. Chem. **1983**, 2151.
41. KELLY, T.R., N.D. PAREKH, and E.N. TRACHTENBERG: Regiochemical Control in the Diels-Alder Reaction of Substituted Naphthoquinones. The Directing Effects of C-6 Oxygen Substituents. J. Org. Chem. **47**, 5009 (1982).
42. BOECKMAN, R.K., JR., T.M. DOLAK, and K.O. CULOS: Diels-Alder Cycloaddition of Juylone Derivatives: Elucidation of Factors Influencing Regiochemical Control. J. Am. Chem. Soc. **100**, 7098 (1978).
43. ROZEBOOM, M.D., L.-M. TEGMO-LARSSON, and K.N. HOUK: J. Org. Chem. **46**, 2338 (1981).
44. SAVARD, J., and P. BRASSARD: Regiospecific Syntheses of Quinones using Vinylketene Acetals Derived from Unsaturated Esters. Tetrahedron Lett. **1979**, 4911.
45. PEARLMAN, B.A., J.M. MCNAMARA, I. HASAN, S. HATAKEYAMA, H. SEKIZAKI, and Y. KISHI: Practical Total Synthesis of ($\pm$)-Aklavinone and Total Synthesis of Aklavin. J. Am. Chem. Soc. **103**, 4248 (1981).
46. GESSON, J.P., J.C. JACQUESY, and B. RENOUX: Approach to the Synthesis of Steffimy-

87. GENOT, A., J.-C. FLORENT, and C. MONNERET: Anthracyclinones. 3. Chiral Pool Synthesis of Anthracyclinones via Tetralin Intermediates. J. Org. Chem. **52**, 1057 (1987).
88. KROHN, K., and H. RIEGER: Synthetische Anthracyclinone XXXIV. Ein enantiomerenreiner AB-Baustein zur Daunomycinon-Synthese unter Einbau von (S)-Äpfelsäure. Liebigs Ann. Chem. **1987**, 517.
89. SEEBACH, D., R. NAEF, and G. CALDERARI: α-Alkylation of α-Heterosubstituted Carboxylic Acids Without Racemization. Tetrahedron **40**, 1313 (1984).
90. COREY, E.J., and M. CHAYKOVSKY: Methylsulfinyl Carbanion (CH_3-SO-CH_2^-). Formation and its Application to Organic Synthesis. J. Am. Chem. Soc. **87**, 1345 (1965).
91. BROADHURST, M., C.H. HASSALL, and G.J. THOMAS: Anthracyclines. Part 2. Investigations relating to the Synthesis of 4-Demethoxyanthracyclinones. J. Chem. Soc. Perkin Trans. 1 **1982**, 2239.
92. KROHN, K., H. RIEGER, E. BROSER, P. SCHIESS, S. CHEN, and T. STRUBIN: Synthese symmetrisch substituierter Daunomycinon-Derivate. Liebigs. Ann. Chem. **1988**, 943.
93. WONG, C.M., R. SCHWENK, D. POPEIN, and T. HO: The Total Synthesis of Daunomycinone. Can. J. Chem. **51**, 466 (1973).
94. LOWN, J.W., S.M. SONDHI, S.B. MANDAL, and J. MURPHY: Synthesis and Redox Properties of Chromophore Modified Glycosides Related to Anthracyclines. J. Org. Chem. **47**, 4304 (1982).
95. TRACY, M., and E.M. ACTON: Synthesis of Tetrahydrobenzo{b}phenazines as Anthracyclinone N-Isosters. J. Org. Chem. **49**, 5116 (1984).
96. BROADHURST, M.J., C.H. HASSALL, and G.J. THOMAS: Total Synthesis of 4-Demethoxydaunomycinone. J. Chem. Soc., Chem. Commun. **1982**, 158.
97. CHENARD, B.L., M.G. DOLSON, A.D. SERCEL, and J.S. SWENTON: Annelation Reactions of Quinone Monoketals. Studies Directed at an Efficient Synthesis of Anthracyclinones. J. Org. Chem. **49**, 318 (1984).
98. SWENTON, J.S., F.K. ANDERSON, D.K. JACKSON, and L.NARASIMHAN: 1,4-Dipole-Metalated Quinone Stratergy to ($\pm$)-4-Demethoxydaunomycinone and ($\pm$)-Daunomycinone. Annelation of Benzocyclobutenedione Monoacetals with Lithioquinone Bisketals. J. Org. Chem. **46**, 4825 (1981).
99. GARDNER, J.N., F.E. CARBON, and D. GNOJ: A One-Step Procedure for the Preparation of Tertiary α-Ketols from the Corresponding Ketones. J. Org. Chem. **33**, 3294 (1968).
100. DEL NERO, S., C. GANDOLFI, P. LOMBARDI, and F. ARCAMONE: Preparation of ($-$)-1,4-Dimethoxy-6-acetyl-6-hydroxytetralin: Intermediate for Anthracyclinones Synthesis. Chem. Ind. (London) **1981**, 810.
101. BROADHURST, M.J., C.H. HASSALL, and G.J. THOMAS: Anthracyclines. Part 3. The Total Synthesis of 4-Demethoxydaunomycin. J. Chem. Soc. Perkin Trans. 1 **1982**, 2249.
102. ––– Total Synthesis of some new 4-Demethoxyanthracyclinones. Tetrahedron **40**, 4649 (1984).
103. JEW, S., S. TERASHIMA, and K. KOGA: Asymmetric Halolactonization Reactions. 3. Asymmetric Synthesis of Optically Active Anthracyclinones. Chem. Pharm. Bull. **27**, 2351 (1979).
104. RUSSELL, R.A., P.S. GEE, R.W. IRVINE, and R.N. WARRENER: Anthracyclines. X. The Enantiospecific Synthesis of ($-$)-(7R)-7-Acetyl-7-hydroxy-4,4-dimethoxy-5,6,7,8-tetrahydronaphthalen-1(4H)-one; a Type I Chiral Dienone for the Synthesis of 7-Deoxydaunomycinone. Aust. J. Chem. **37**, 1709 (1984).
105. TOMIOKA, K., M. NAKAJIMA, and K. KOGA: Enantioface Differentiation in Cis Dihydroxylation of C-C Double Bonds by Osmium Tetroxide with Use of a Chiral Diamine with D_2 Symmetry. J. Am. Chem. Soc. **109**, 6213 (1987).

106. MOORE, J.A., and M. RAHM: The Synthesis of 6-Hydroxy-1,3,4,5-tetrahydrobenz-[cd]indole. J. Org. Chem. **26**, 1109 (1961).
107. WATANABE, M., H. MAENOSONO, and S. FURUKAWA: A Regiospecific Synthesis of Anthracyclinones Using directed Metalation. Chem. Pharm. Bull. **31**, 2662 (1983).
108. BLADE, R.J., and P. HODGE: Synthetic Routes to ($\pm$)-Daunomycinone: Elaboration of the Hydroxy-ketone Group from an α-Tetralone Derivative, and Selective Methylation of C(4)-Hydroxy Group Using Diazomethane. J. Chem. Soc., Chem. Commun. **1979**, 85.
109. FRESKOS, J.N., G.W. MORROW, and J.S. SWENTON: Synthesis of Functionalized Hydroxyphthalides and Their Conversion to 3-Cyano-1(3H)-isobenzofuranones. The Diels-Alder Reaction of Methyl 4,4-Diethoxybutynoate and Cyclohexadienes. J. Org. Chem. **50**, 805 (1985).
110. RUSSELL, R.A., G.J. COLLIN, M.P. CRANE, P.S. GEE, A.S. KRAUSS, and R.N. WARRENER: Anthracyclines. XI. A Short, Site-Specific Synthesis of Unsymmetrical 3-Acetyl-5,8-dialkoxy-1,2-dihydronaphthalenes; Key Precursors to Daunomycinone AB-Synthons. Aust. J. Chem. **37**, 1721 (1984).
111. REDDY, M.P., and G.S.K. RAO: A High Yield Synthesis of ($\pm$)-5,8-Dimethoxy-2-α-hydroxymethyl-3,4-dihydronaphthalene, a Key Intermediate in Anthracyclinone Synthesis. Tetrahedron Lett. **22**, 3549 (1981).
112. RAO, A.V.R., J.S. YADAV, K.B. REDDY, and A.R. MEHENDALE: A Stereoconvergent Synthesis of (+)-4-Demethoxydaunomycinone. J. Chem. Soc., Chem. Commun. **1984**, 453.
113. ---- A Stereoconvergent Synthesis of (+)-4-Demethoxydaunomycinone. Tetrahedron **40**, 4643 (1984).
114. RAO, A.V.R., B.M. CHANDRA, and H.B. BORATE: A General Method for the Synthesis of 11-Deoxyanthracyclinones. Synth. Commun. **14**, 257 (1984).
115. RAO, A.V.R., K.B. REDDY, and A.R. MEHENDALE: A Regiospecific and Flexible Approach for the Synthesis of ($\pm$)-Daunomycinone and ($\pm$)-11-Deoxydaunomycinone. J. Chem. Soc., Chem. Commun. **1983**, 564.
116. RAO, A.V.R., V.H. DESHPANDE, and N.L. REDDY: A Short Synthesis of ($\pm$)-2-Acetyl-5,8-dimethoxy-1,2,3,4-tetrahydro-2-naphthol – A Key Intermediate for Anthracyclinone Synthesis. Tetrahedron Lett. **23**, 4373 (1982).
117. CAMBIE, R.C., T.A. HOWE, M.G. PAULSER, P.S. RUTLEDGE, and P.D. WOODGATE: Experiments directed towards the Synthesis of Anthracyclines. XIV. An Intermediate for the Synthesis of Vineomycins. Aust. J. Chem. **40**, 1063 (1987).
118. CAMBIE, R.C., D.S. LARSEN, P.S. RUTLEDGE, and P.D. WOODGATE: Experiments Directed Towards the Synthesis of Anthracyclinones. XIII. Ozonolysis of Anthrafurans. Aust. J. Chem. **40**, 215 (1987).
119. COBURN, C.E., D.K. ANDERSON, and J.S. SWENTON: Convenient AB-Ring Segments for Anthracyclinone Synthesis via Bishydroxylation of 2-Ethyl-5,8-dimethoxy-7-bromo-1-tetralone. J. Org. Chem. **48**, 1455 (1983).
120. SWENTON, J.S., D.K. ANDERSON, C.E. COBURN, and A.P. HAAG: The Synthesis of 6-Demethoxyanthracyclinones: ($\pm$)-Citromycinone and ($\pm$)-Demethoxy-deoxydaunomycinone. Tetrahedron **40**, 4633 (1984).
121. LI, T.-T., Y.L. WU, and T.C. WALSGROVE: A Facile Total Synthesis of Racemic Aklavinone. Tetrahedron **40**, 4701 (1984).
122. HAUSER, F.M., and D. MAL: Total Synthesis of 11-Deoxydaunomycinone. J. Am. Chem. Soc **105**, 5688 (1983).
123. RAO, A.V.R., V.H. DESHPANDE, and N.L. REDDY: A Simple Synthesis of ($\pm$)-4-Demethoxydaunomycinone. Tetrahedron Lett. **21**, 2661 (1980).
124. KENDE, A.S., D.P. CURRAN, Y. TSAY, and J.E. MILLS: The Isobenzofurane Route to Anthracyclinones. Tetrahedron Lett. **1977**, 3537.

125. UMEZAWA, H., Y. TAKAHASHI, M. KINOSHITA, H. NAGANAWA, K. TATSUTA, and T. TAKEUCHI: Synthesis of 4-Demethoxy-11-deoxy-analogs of Daunomycin and Adriamycin. J. Antibiotics **33**, 1581 (1980).
126. DESHPANDE, V.H., K. RAVICHANDRAN, and B.R. RAO: Synthesis of ($\pm$)-4-Demethoxy-7,11-dideoxydaunomycinone. Synth. Commun. **14**, 477 (1984).
127. SMITH, T.H., A.N. FUJIWARA, W.W. LEE, H.Y. WU, and D.W. HENRY: Synthetic Approaches to Adriamycin. 2. Degradation of Daunorubicin to a Nonasymmetric Tetracyclic Ketone and Refunctionalization to the A Ring to Adriamycin. J. Org. Chem. **42**, 3653 (1977).
128. ALEXANDER, J., I. KHANNA, D. LEDNICER, L.A. MITSCHER, T. VEYSOGLU, Z. WIELOGORSKI, and R.L. WOLGEMUTH: Total Chemical Synthesis and Antitumor Evaluation of 4-Demethoxy-10,10-dimethyldaunomycinone. J. Med. Chem. **27**, 1343 (1984).
129. WISEMAN, J.R., N.I. FRENCH, R.K. HALLMARK, and K.G. CHIONG: The *ortho*-Quinodimethane Route to Anthracyclinones. A New Synthesis of 4-Demethoxydaunomycinone. Tetrahedron Lett. **1978**, 3765.
130. PRESTON, P.N., T. WINWICK, and J.O. MORLEY: Selective Demethylation of Di- and Trimethoxyanthraquinones via Aryloxydifluoroboron chelates. Synthesis of 4-Hydroxy-1,5-dimethoxyanthraquinone and 1,4-Dihydroxy-5-methoxyanthraquinone. J. Chem. Soc., Chem. Commun. **1983**, 89.
131. SUZUKI, M., Y. KIMURA, and S. TERASHIMA: A Novel Synthesis of the α-Hydroxyketone Moiety of Anthracyclinones by the Use of 2-Trimethylsilylethynylcerium(III) Reagents. Chem. Lett. **1984**, 1543.
132. TAMURA, Y., M. SASHO, H. OHE, S. AKAI, and Y. KITA: A Short and Efficient Synthesis of 11-Deoxyanthracyclinones: Strong Base-Induced Cycloaddition of the Suitably Substituted Tetrahydrohomophthalic Anhydride. Tetrahedron Lett. **26**, 1549 (1985).
133. SUZUKI, M., Y. KIMURA, and S. TERASHIMA: Novel Ethynylcerium(III) Reagents as Effective Tools for Constructing the α-Hydroxy Methyl Ketone Moiety of Anthracyclinones. Chem. Pharm. Bull. **34**, 1531 (1986).
134. TAMURA, Y., M. SASHO, S. AKAI, and H. KISHIMOTO: A Highly Convergent Strategy for the Synthesis of 4-Demethoxydaunomycinone and Daunomycinone: A Novel Synthesis of C4-Acetoxylated Homophthalic Anhydrides. Tetrahedron Lett. **27**, 195 (1986).
135. TAMURA, Y., S. AKAI, H. KISHIMOTO, M. KIRIHARA, M. SASHO, and Y. KITA: Practical Total Synthesis of 11-Deoxydaunomycinone and the First Total Synthesis of 11-Deoxydaunomycin. Tetrahedron Lett. **28**, 4583 (1987).
136. SODEOKA, M., T. IIMORI, and M. SHIBASSAKI: Stereospecific Synthesis of exo-Allylic Alcohol. An Efficient Asymmetric Synthesis of (R)-(−)-2-Acetyl-5,8-dimethoxy-1,2,3,4-tetrahydro-2-naphthol. Tetrahedron Lett. **26**, 6497 (1985).
137. HIYAMA, T., M. SAWAHATA, and Y. KUSANO: Manganese-Mediated Allyl Addition to Enolizable Aldehyde and Ketones. An Approach for Introduction of Acetonyl Side Chains at C(9) of Anthracycline Antibiotics. Chem. Lett. **1985**, 611.
138. ISHIZUMI, K., N. OHASHI, and N. TANNO: Stereospecific Total Synthesis of 9-Aminoanthracyclines: (+)-9-Amino-9-deoxydaunomycin and Related Compounds. J. Org. Chem. **52**, 4477 (1987).
139. TAMURA, Y., H. ANNOURA, H. YAMAMOTO, H. KONDO, Y. KITA, and H. FUJIOKA: Asymmetric Synthesis of Anthracyclinones Using Chiral Acetal: Synthesis of a new Chiral AB-Synthon, (−)-2-Bromo-6-ethynyl-6-hydroxy-5,6,7,8-tetrahydro-1,4-naphthoquinone, and its Application for (−)-7-Deoxydaunomycinone. Tetrahedron Lett. **28**, 5709 (1987).
140. ARDECKY, R.J., D. DOMINGUEZ, and M.P. CAVA: 3-(Acyloxy)-3-buten-2-ones as Dienophiles in Anthracyclinone Synthesis. An Efficient Route to 4-Demethoxy-7-deoxydaunomycinone Derivatives. J. Org. Chem. **47**, 409 (1982).

141. STEPHAN, D., A. GORGUES, and A. LE COQ: Reduction D'α,α-Dibromoorthoxylenes par les sels Chromeux: Génération Facile D'Orthoquinodimethanes. Tetradedron Lett. **25**, 5649 (1984).
142. ARDECKY, R.J., F.A.J. KERDESKY, and M.P. CAVA: Dienophilic Reactions of 3-[(Trimethylsilyl)oxy]-3-buten-2-one. J. Org. Chem. **46**, 1483 (1981).
143. HONEK, J.F., M.L. MANCINI, and B. BELLEAU: New Strategy for the Synthesis of Key Anthracycline Precursors. Tetrahedron Lett. **24**, 257 (1983).
144. KROHN, K.: Synthetische Anthracyclinone, XVIII. Synthese des 13-Desoxo-6-desoxydaunomycinons und des β_1-Citromycinons. Liebigs Ann. Chem. **1981**, 2285.
145. KERDESKY, F.A.J., R.J. ARDECKY, M.V. LAKSHMIKANTHAM, and M.P. CAVA: Simple o-Quinodimethane Route to ($\pm$)-4-Demethoxydaunomycinone. J. Am. Chem. Soc. **103**, 1992 (1981).
146. WATABE, T., Y. TAKAHASHI, and M. ODA: 1,2-Dihydro-3,10-dihydroxycyclobut-[b]anthracene-4,9-dione; a Key Intermediate for 4-Demethoxyanthracyclinones. Tetrahedron Lett. **24**, 56 (1983).
147. KEAY, B.A., and R. RODRIGO: A Convergent Synthesis of ($\pm$)-Daunomycinone. Tetrahedron **40**, 4597 (1984).
148. SIBI, M.P., N. ALTINTAS, ans V. SCIECKUS: Benzamide Directed Ortho Metalation. A Route to the A/B Ring Synthon of Daunomycinone. Tetrahedron **40**, 4593 (1984).
149. RUSSELL, R.A., A.S. KRAUSS, R.W. IRVINE, and R.N. WARRENER: Anthracyclines. XII. The Preparation of ($-$)-(7R)-7-Acetyl-7-(t-butyldimethylsiloxy)-4,4-dimethoxy-5,6,7,8-tetrahydronaphthalene-1(4H)-one and ($+$)-(5R)-6-(t-butyldimethylsiloxy)-4.4-dimethoxy-5,6,7,8-tetrahydronaphthalene-1(4H)-one: an Improved Route to Chiral AB Synthons for 7-Deoxydaunomycinone. Aust. J. Chem. **38**, 179 (1985).
150. IRVINE, R.W., R.A. RUSSELL, and R.N. WARRENER: Allylic Vs Homoallylic Control of Stereospecifity in the Epoxidation of 3(1'-Hydroxyethyl)-5,8-dimethoxy-1,2-dihydronaphthalen-1-ol: Implications for the Synthesis of Chiral Anthracyclines. Tetrahedron Lett. **26**, 6117 (1985).
151. TAMOTO, K., and S. TERASHIMA: Novel Synthesis of Optically Pure Anthracyclinone Intermediates by the Use of Microbial Asymmetric Reduction with Fermenting Baker's Yeast. Chem. Pharm. Bull. **32**, 4328 (1984).
152. SUZUKI, M., Y. KIMURA, and S. TERASHIMA: A Novel and Highly Efficient Asymmetric Synthesis of Optically Active Anthracyclinones. Bull. Chem. Soc. Jpn. **39**, 3559 (1986).
153. BAUMAN, J.G., R.C. HAWLEY, and H. RAPOPORT: An Efficient Synthesis of Aklavinone and Related 11-Deoxyanthracyclinones. J. Org. Chem. **50**, 1569 (1985).
154. ——— Synthesis of a Cyclohexadiene Monoepoxide by Intramolecular Darzens Condensation. Efficient Synthesis of an A-Ring Anthracyclinone Precursor. J. Org. Chem. **49**, 3791 (1984).
155. DIMROTH, O., and T. FAUST: Über die Borsäureester der Oxyanthrachinone. Ber. dtsch. chem. Ges. **54**, 3020 (1921).
156. MEYERS, A.I., and K. HIGASHIYAMA: Asymmetric Additions to Chiral Naphthalenes. 4. An Asymmetric Synthesis of the AB-Ring of Aklavinone. J. Org. Chem. **52**, 4592 (1987).
157. PENCO, S., F. ANGELUCCI, M. BALLABIO, G. BARCHELLI, A. SUARATO, E. VANOTTI, A. VIGEVANI, and F. ARCAMONE: Regiospecific Total Synthesis of 6-Deoxyanthracyclines. Tetrahedron **40**, 4677 (1984).
158. ——————— Synthesis and Ring Conformation of New Anthracyclines. Heterocycles **21**, 21 (1984).
159. ANGELUCCI, F., G. BARCHIELLI, G.A. BRUSSANI, M. GIGLI, B. GIOIA, R. HERMANN, A. SUARATO, E. VANOTTI, and S. PENCO: Flexible Synthetic Route to 6-Deoxy and 11-Deoxyanthracyclinones. Tetrahedron Lett. **26**, 5693 (1985).

160. KELLY, T.R., and W.-G. TSANG: The Synthesis of 4-Demethoxydaunomycinone. Tetrahedron Lett. **1978**, 4457.
161. LEE, W.W., A.P. MARTINEZ, T.H. SMITH, and D.W. HENRY: Daunomycinone Analogues via the Diels-Alder Reaction. Synthesis and Chemistry of Some 6,11-Dihydroxy-5,12-naphthacenediones. J. Org. Chem. **41**, 2296 (1976).
162. VAN KLEEF, R.P., F.J., POTMAN, and H.W. SCHEEREN: Tailor-Made Butadienes for the Site-Selective Cycloaddition with Quinizarinquinone and Other Unsymmetrically Substituted Quinones. J. Org. Chem. **50**, 1955 (1985).
163. KENDE, A.S., Y. TSAY, and J.E. MILLS: Total Synthesis of ($\pm$)-Daunomycinone and ($\pm$)-Carminomycinone. J. Am. Chem. Soc. **98**, 1967 (1976).
164. KIMURA, Y., M. SUZUKI, T. MATSUMOTO, R. ABE, and S. TERASHIMA: 2-Chloro-1,3-butadiene as an Efficient Enophile in the Diels-Alder Reaction with Anthracene-1,4,9,10-tetraone. A Simple Synthesis of the 4-Demethoxyanthracyclinone Intermediate. Chem. Lett. **1984**, 473.
165. SMITH, T.H., A.N. FUJIWARA, D.W. HENRY, and W.W. LEE: Synthetic Approaches to Adriamycin. Degradation of Daunorubicin to Nonasymmetric Tetracyclic Ketone and Refunctionalization of the A-ring to Adriamycin. J. Am. Chem. Soc. **97**, 1969 (1976).
166. SUZUKI, M., T. MATSUMOTO, R. ABE, Y. KIMURA, and S. TERASHIMA: A Simple and Efficient Synthesis of Key Synthetic Intermediates of 4-Demethoxyanthracyclinones, ($\pm$)- and (R)-(−)-7-Deoxy-4-demethoxydaunomycinone. Chem. Lett. **1985**, 57.
167. GESSON, J.P., J.C. JACQUESY, and M. MONDON: Cycloaddition d'acetals de céténes silylés avec diverses naphthoquinones. Méthode générale de synthèse d'intermediates en série anthracycline. Nouv. J. Chim. **7**, 205 (1983).
168. PRESTON, P.N., and T. WINWICK: Synthesis of Intermediates related to 11-Deoxyanthracyclinones. J. Chem. Soc. Perkin Trans. 1 **1985**, 39.
169. BAUMAN, J.G., R.B. BARBER, R.D. GLESS, and H. RAPOPORT: The Vinylketene Acetale Route to Aklavinone and 11-Deoxydaunomycinone. Tetrahedron Lett. **21**, 4777 (1980).
170. TAMURA, Y., A. WADA, M. SASHO, K. FUKUNAGA, H. MAEDA, and Y. KITA: A New General Regiocontrolled Synthesis of Anthracyclinones Using Cycloaddition of Homophthalic Anhydrides to 2-Chloro-6-oxo-5,6,7,8-tetrahydro-1,4-naphthoquinone 1,2-Ethanediyl Acetal. J. Org. Chem. **47**, 4376 (1982).
171. VEDEJS, E., W.H. MILLER, and J.R. PRIBISH: Silicon-Mediated Synthesis of 11-Deoxyanthracyclines. J. Org. Chem. **48**, 3613 (1983).
172. TAMURA, Y., M. SASHO, S. AKAI, A. WADA, and Y. KITA: Anthracyclinone Synthesis Using Strong Base Induced Cycloaddition of Homophthalic Anhydrides and Related Compounds. Tetrahedron **40**, 4539 (1984).
173. KROHN, K., and W. BALTUS: Synthesis of Daunomycinone via Intramolecular Nucleophilic Addition to 9,10-Anthraquinones. Synthesis **1986**, 942.
174. KHANAPURE, S.P., R.T. REDDY, and E.R. BIEHL: The Preparation of Anthaquinones and Anthracyclinones via the Reaction of Haloarenes and Cyanophthalides under Aryne-Forming Conditions. J. Org. Chem. **52**, 5685 (1987).
175. FARINA, F., T. MOLINA, and M.C. PAREDES: Synthesis, Tautomerism and Diels-Alder Reactions of 1,4-Dihydroxy-9,10-anthraquinon-9-imines. Tetrahedron Lett. **26**, 111 (1985).
176. RUSSELL, R.A., G.J. COLLIN, M. STERNS, and R.N. WARRENER: Cycloaddition Routes to Polycyclic Quinones: Part 1; Boron Triacetate as a Regiochemical Directing Agent. Tetrahedron Lett. **1979**, 4229.
177. RAO, A.V.R., G. VENKATSWAMY, S.M. JAVEED, V.H. DESHPANDE, and B.R. RAO: Synthesis of ($\pm$)-4-Demethoxydaunomycinone. J. Org. Chem. **48**, 1552 (1983).

178. GUPTA, D.N., P. HODGE, and N. KHAN: Chemistry of Quinones. Part 7. Synthesis of Anthracyclinone Anologues via Diels-Alder Reactions of 1,4-Anthraquinones. J. Chem. Soc. Perkin Trans. 1 **1981**, 689.
179. FARINA, F., and P. PRADOS: Synthesis of Tetracyclic Hydroxyketones Related to Daunomycinone. Tetrahedron Lett. **1979**, 477.
180. KROHN, K., and K. TOLKIEHN: Synthetische Anthracyclinone, VIII. Totalsynthese des Daunomycinons. Chem. Ber. **112**, 3453 (1979).
181. — — Synthetische Anthracyclinone, XIV. Synthese neuer Derivate des Daunomycinons und des β-Rhodomycinons. Chem. Ber. **113**, 2976 (1980).
182. KROHN, K., K. TOLKIEHN, V. LEHNE, H.W. SCHMALLE, and H.-F. GRÜTZMACHER: Synthetic Anthracyclinones, XXIX. Quinone Antibiotics with Five Substituents at the Hydroaromatic Ring. Liebigs Ann. Chem. **1985**, 1311.
183. TAMIREZ, J., and P. VOGEL: A Doubly Convergent and Regioselective Synthesis of ($\pm$)-Daunomycinone. Tetrahedron **40**, 4549 (1984).
184. GUPTA, R.C., P.A. HARLAND, and R.J. STOODLEY: An Efficient Enantiocontrolled Synthesis of (+)-4-Demethoxydaunomycinone. Tetrahedron **40**, 4657 (1984).
185. GUPTA, R.C., D.A. JACKSON, R.J. STOODLEY, and D.J. WILLIAMS: Studies Related to Anthracyclinones. Part 2. Synthesis of ($\pm$)-4-Demethoxydaunomycinone. J. Chem. Soc. Perkin Trans. 1 **1985**, 525.
186. TROST, B.M., D. O'KRONGLY, and J.L. BELLETIRE: A Model for Asymmetric Induction in the Diels-Alder Reaction. J. Am. Chem. Soc. **102**, 7595 (1980).
187. KROHN, K., and W. BALTUS: Synthesis of *rac-* and *ent-*Fridamycin E. Tetrahedron **44**, 49 (1988).
188. KROHN, K., and B. SARSTED: Facile Synthesis of the ($\pm$)-6-Deoxycarminomycinone from Aloe-emodine. Angew. Chem. **95**, 897 (1983); Angew. Chem. Int. Ed. **22**, 875 (1983).
189. BENNANI, F., J.-C. FLORENT, M. KOCH, and C. MONNERET: An Efficient Synthesis of Optically active 4-Demethoxy Anthracyclinones. Tetrahedron **40**, 4669 (1984).
190. MINCHER, D.J., G. SHAW, and E. DE CLERCQ: Anthracyclinones. Part 1. A Versatile Synthesis of the Anthracyclinone System using a Chiral Template derived from a Carbohydrate. J. Chem. Soc. Perkin Trans. I **1983**, 613.
191. KROHN, K., U. MÜLLER, W. PRIYONO, B. SARSTEDT, and A. STOFFREGEN: Intramolekulare Addition von Carbanionen an Anthrachinone. Liebigs Ann. Chem. **1984**, 306.
192. KROHN, K., and W. PRIYONO: Simple Synthesis of Anthracyclinones by Cyclization of an Intermediate Hydroxynitronate. Angew. Chem. **98**, 338 (1986); Angew. Chem. Int. Ed. Engl. **25**, 339 (1986).
193. CAVA, M.P., Z. AHMED, N. BENFAREMO, R.A. MURPHY JR., and G.J. O'MALLEY: Anthraquinone Dye Intermediates as Precursors of Aklavinone-Type Anthracyclinones. Tetrahedron **40**, 4767 (1984).
194. MCNAMARA, J.M., and Y. KISHI: Practical Asymmetric Synthesis of Aklavinone. Tetrahedron **40**, 4685 (1984).
195. KROHN, K., M. KLIMARS, H.J. KÖHLE, and E. EBELING: Synthesis of ζ-Pyrromycinon, 7-Deoxyauramycinone, and 7-Deoxyaklavinone via Ketoester Cyclization. Tetrahedron **40**, 3677 (1984).
196. GESSON, J.-P., J.-P. JACQUESY, and B. RENOUX: A General and Regiospecific Route to Tetracyclic Alkenes in the 11-Deoxyanthracyclinone Series. Application to the Total Synthesis of ($\pm$)-Auramycinone. Tetrahedron **40**, 4743 (1984).
197. SNIECKUS, V.: New Directions in Heterocyclic Synthesis using Metalated Benzamides. J. Heterocycl. Chem. **7**, 95 (1984).
198. TAUB, D., N.N. GIROTRA, R.D. HOFFSOMMER, C.H. KUO, H.L. SLATES, S. WEBER, and N.L. WENDLER: Total Synthesis of the Macrolide Zearalenone. Tetrahedron **24**, 2443 (1968).

199. RUSSELL, R.A., B.A. PILLEY, R.W. IRVINE, and R.N. WARRENER: Anthracyclines, XVI. Further Comments Concerning the Phthalide Anion Annelation of Quinone Monoacetals. Aust. J. Chem. **40**, 311 (1987).
200. RUSSELL, R.A., B.A. PILLEY, and R.N. WARRENER: A High-Yielding Synthesis of 3-Cyanophthalides. Synth. Commun. **16**, 425 (1986).
201. RUSSELL, R.A., R.W. IRVINE, and R.N. WARRENER: The Total Synthesis of Optically Pure (9R,13S)- and (9R,13R)-7-Deoxy-13-dihydrodaunomycinone. J. Org. Chem. **51**, 1595 (1986).
202. RUSSELL, R.A., A.S. KRAUSS, R.N. WARRENER, and R.W. IRVINE: A High-yielding Enantiospecific Synthesis of (−)-7-Deoxydaunomycinone (Part 9). Tetrahedron Lett. **25**, 1517 (1984).
203. HAUSER, F.M., and V.M. BAGHDANOV: Regiospecific Total Synthesis of (±)-Daunomycinone from an 11-Deoxydaunomycinone Precursor. Tetrahedron **40**, 4719 (1984).
204. JUNG, M.E., and J.A. LOWE: Synthetic Approaches to Adriamycin Involving Diels-Alder Reactions of Photochemically generated Bisketenes. Total Synthesis of Islandicin and Digitopurpone. J. Org. Chem. **42**, 2371 (1977).
205. NOMURA, K., K. OKAZAKI, K. HORI, and E. YOAHII: Total Synthesis of (±)-Granaticin. J. Am. Chem. Soc. **109**, 3402 (1987).
206. HAUSER, F.M., R.P. RHEE, S. PRASANNA, S.M., WEINREB, and J.H. DODD: Ortho-Toluate Carbanion Chemistry: Sulfenylation and Selenation. Synthesis **1980**, 72.
207. PARKER, K.A., D.M. SPERO, and K.A. KOZISKI: Evaluation of Some Preparations of Trialkoxyphthalic Acid Derivatives. J. Am. Chem. Soc. **52**, 183 (1987).
208. BIRCH, A.J., and K.P. DASTUR: A Catalytic Conversion of 1-Methoxycyclohexa-1,4-dienes into 1-Methoxycyclohexa-1,3-dienes. Tetrahedron Lett. **1972**, 4195.
209. HARLAND, P.A., and J.H. DODD: Synthesis of Phthalates, Benzoates, and Phthalides via the in situ Generation of Methoxycyclohexa-1,3-dienes and their Subsequent Diels-Alder Reactions with Acetylenes. Synthesis **1982**, 223.
210. JOHNSON, O.H., and J.R. HOLUM: 1,1-Diethoxy-3-(triphenylstannyl)-2-propyne. J. Org. Chem. **23**, 738 (1958).
210a. TAMURA, Y., M. SASHO, S.AKAI, H. KISHIMOTO, J. SEKIHACHI, and Y. KITA: An Efficient, Regiospecific Synthesis of 4-Demethoxydaunomycinone and Daunomycinone. Chem. Pharm. Bull. **35**, 1405 (1987).
211. RAVICHANDRAN, K., F.A.J. KERDESKY, and M.P. CAVA: Synthesis of 4-Demethoxy-6,11-dideoxydaunomycinone. A Highly Deoxygenated Anthracyclinone. J. Org. Chem. **51**, 2044 (1986).
212. VEDEJS, E.V., and J.R. PRIBISH: A Synthesis of the C_6,C_{11}-Dideoxyanthracyclinone Skeleton via Hassall Cyclization and Oxidative Desililation. J. Org. Chem. **53**, 1593 (1988).
213. MATSUMOTO, T., M. OHSAKI, F. MATSUDA, and S. TERASHIMA: Effecient Synthesis and Antitumor Activity of Novel 14-Fluoroanthracyclines. Tetrahedron Lett. **28**, 4419 (1987).
214. LOWN, J.W., and S.M. SONDHI: Glycosidic Coupling of Regiospecifically Synthesized Xantho[2,3-g]tetralin Aglycones to Afford Moderately Antileukemic but Redox Inactive Structures Related to Anthracyclines. J. Org. Chem. **50**, 1413 (1985).
215. WONG C.-M., W. HAQUE, H.-Y. LAM, K. MARAT, and E. BOCK: Heteroanthracyclines. 1,4-Demethoxyxanthodaunomycinone (6,7,9,11-tetrahydroxy-9-acetyl-7,8,9,10-tetrahydrobenzo(B)xanthen-12-one). Can. J. Chem. **61**, 1788 (1983).
216. KENDE, A.S., and H. NEWMANN: Eur. Pat. Appl. (1980); CA: **94**, 140117j (1981).
217. KAWASAKI, M., F. MATSUDA, and T. TERASHIMA: First Total Synthesis of (+)-7-Deoxynogarol and (+)-7-Con-O-methylnogarol. Tetrahedron Lett. **29**, 791 (1988).

(*Received April 1, 1988*)

Indole Alkaloid Production in Catharanthus roseus Cell Suspension Cultures

By M. LOUNASMAA and J. GALAMBOS[1], Laboratory for Organic and Bioorganic Chemistry, Technical University of Helsinki, Finland

Contents

1. Introduction 89
2. Discussion 90
   2.1. Indole Alkaloids and Their Formation in the Plant and in Cell Cultures . . . 90
   2.2. General Methods of *Catharanthus roseus* Cell Suspension Culture Work . . . 97
      2.2.1. Analytical Methods 97
      2.2.2. Development of High Yielding Cell Lines 98
      2.2.3. Growth and Alkaloid Production in *Catharanthus roseus* Cell Suspension Cultures 99
      2.2.4. Effects of Culture Conditions on Growth and Alkaloid Production . . 101
         2.2.4.1. Medium Composition 101
         2.2.4.2. Addition of Precursors 103
         2.2.4.3. Light 104
         2.2.4.4. Temperature 104
         2.2.4.5. Gaseous Environment 105
      2.2.5. One-stage Systems 105
      2.2.6. Large-scale Fermentation 105
      2.2.7. Immobilized Cell Systems 106
      2.2.8. Cell Free Systems 106
3. Conclusions 107
References 108

1. Introduction

Catharanthus roseus (L.) G. Don, Madagascan periwinkle, is a well-known medicinal herb on which a lot of research has been done, especially on isolation and conversion of its alkaloids. To date more than 100 indole alkaloids have been isolated from the different parts

---

[1] Permanent address: Gedeon Richter Ltd, Budapest, Hungary

of the plant, many of them with important pharmacological activity (*1–3*). The most notable of these therapeutic alkaloids are the antileukemic drugs vinblastine (**1**) and vincristine (**2**), the antihypertensive ajmalicine (**3**) used in combination with reserpine, and serpentine (**4**) which has a sedative effect. The generally low concentration of these products in a plant which is itself scarce, combined with the difficulty of separating the valuable alkaloids from others co-occurring ones, has encouraged intensive research on their synthesis or semisynthesis (*4–11*). So far, however, attempts to produce these alkaloids economically by the normal methods of synthetic organic chemistry have not given satisfactory results.

Recent advances in cell culture techniques and the successful application of the method to several other plant species have turned the attention of researchers to this new approach. Worldwide, attempts are today being made to produce the alkaloids of *C. roseus* by cell culture methods. Already a number of monomeric indole alkaloids have been isolated and some cell lines have been found to produce alkaloids in excess of the levels found in the plant. Although the main target compounds, the dimeric alkaloids vinblastine (**1**) and vincristine (**2**), have so far been detected only in callus or organ cultures of *C. roseus* (*12–13*), it is not unreasonable to expect that as a result of further experimentation cell suspension cultures might one day be the most economical source of these chemotherapeutic agents.

Our aim in preparing this article has been to give a general review of cell culture work with *C. roseus* and critically to summarize the present state of the art, while paying special attention to the problems and results relevant to eventual economical production of the desired indole alkaloids. The time period covered extends from 1975 to August 1988. Numbering of the compounds is based on the biogenetic numbering proposed by LE MEN and TAYLOR (*14*).

2. Discussion

2.1. Indole Alkaloids and Their Formation in the Plant and in Cell Cultures

More than 100 indole alkaloids have been isolated from the different parts of *C. roseus*, including the alkaloid derivatives encountered during biosynthetic studies on seedlings and very young plants. While most of these are monomeric alkaloids belonging to the corynanthe, strychnos, aspidosperma and iboga families, there are also dimeric alkaloids,

References, pp. 108–115

1 R = CH₃
2 R = CHO

among them the two clinically most valuable alkaloids, vinblastine (**1**) and vincristine (**2**).

The pattern of alkaloids found in *C. roseus* cell suspension cultures is simpler. Only 43 monomeric indole alkaloids have been mentioned in references *15–28*[1]. These are ajmalicine (**3**), serpentine (**4**), tetrahydroalstonine (**5**), 3-iso-19-epiajmalicine (**6**), 3-isoajmalicine (**7**), akuammigine (**8**), 7-hydroxyindolenineajmalicine (**9**), pseudoindoxylajmalicine (**10**), mitraphylline (**11**), strictosidine (**12**), sitsirikine (**13**), isositsirikine (16*R*, 19*E*) (**14**), isositsirikine (16*R*, 19*Z*) (**15**), dihydrositsirikine (**16**), yohimbine (**17**), akuammiline (**18**), desacetylakuammiline (**19**), 10-hydroxydesacetylakuammiline (**20**), pleiocarpamine (**21**), 21-hydroxycyclolochnerine (**22**), vindolinine (**23**), 19-epivindolinine (**24**), vindolinine-N_b-oxide (**25**), 19-epivindolinine-N_b-oxide (**26**), tabersonine (**27**), 19-hydroxytabersonine (**28**), 19-acetoxy-11-hydroxytabersonine (**29**), 19-acetoxy-11-methoxytabersonine (**30**), 19-hydroxy-11-methoxytabersonine (**31**), minovincinine (**32**), hörhammericine (**33**), hörhammerinine (**34**), lochenericine (**35**), lochnerinine (**36**), vindoline (**37**), catharanthine (**38**), akuammicine (**39**), vinervine (**40**), strictosidine lactam (**41**), antirhine (**42**), vallesiachotamine (**43**), 17-isovallesiachotamine (**44**) and N,N-dimethyltryptamine (**45**).

Among these 43 monomers, vindoline (**37**) and catharanthine (**38**) are of particular interest as the biosynthetic precursors of the dimeric alkaloids vinblastine (**1**) and vincristine (**2**) (*29, 30*).

A number of the alkaloids belong to the earlier stages of indole alkaloid biogenesis (*29, 30*) and 26 of them (**6–11, 15, 17–20, 22, 24, 26, 29–34, 40–45**) have been isolated only from cell cultures of *C. roseus*.

---

[1] Some erroneous names and formulae appearing in the original articles are corrected in the list that follows.

References, pp. 108–115

Indole Alkaloid Production in *Catharanthus roseus* Cell Suspension Cultures   93

References, pp. 108–115

The finding of so many compounds only in cell cultures suggests that the complex enzymatic system and the environmental conditions of the cell cultures differ from those of the intact *C. roseus* plant and that some steps of the biosynthetic pathway are blocked in the cultured undifferentiated cells.

So far, only few positive results concerning the firm detection of the dimeric alkaloids vinblastine (**1**) and/or vincristine (**2**) in cell cultures in general have been described (*12, 13*) (*vide infra*).

Not only do the alkaloid spectra of cell cultures differ from the spectrum of the intact plant but they differ from each other, too. When *C. roseus* cells are cultured under identical conditions, different spectra are observed in cell lines of different origin, even in cell lines of the same origin. This vividly demonstrates the wide variability of cell lines. However, the occurrence of combinations is not random and certain combinations appear more frequently than others (*17*).

The biogenetic pathway for *Catharanthus* alkaloids has been a matter of much controversy (*29* and references therein). Formerly, the pathway of indole alkaloid formation was investigated *in vivo* by tracer methods. The disadvantage of this method is that feeding of real or proposed precursors to the intact plant does not always result in their incorporation, since because of the very low concentrations they may be degraded by the plant's enzymatic system (*31*). Thus even well-planned experiments may end in failure. Today, the rapid development of cell culture techniques and of cell free systems in combination with new analytical methods (*32*) provides an alternative approach to the investigation of alkaloid biosynthesis; as a result, a considerable part of the biogenetic pathway and the enzymes involved in it is now known. Though different results are sometimes obtained, these differences can usually be explained in terms of variations in the cell lines (*33*).

There is general agreement that the first step in the indole alkaloid pathway in the plant is the enzymatic stereospecific condensation of tryptamine (**46**) with the monoterpenoid iridoid secologanin (**47**), which gives rise to the glucoalkaloid strictosidine (isovincoside) (**12**) (*31, 34*). The catalysing enzyme, strictosidine synthase (*35*), has been purified and its role in the biosynthesis confirmed (*36, 37*). The next step is the hydrolysis of strictosidine (**12**) and then, in several steps which are not entirely clear (*33, 35*), the formation of the corynanthe-type alkaloids ajmalicine (**3**), tetrahydroalstonine (**5**) and 19-epiajmalicine (**48**) (Scheme 1).

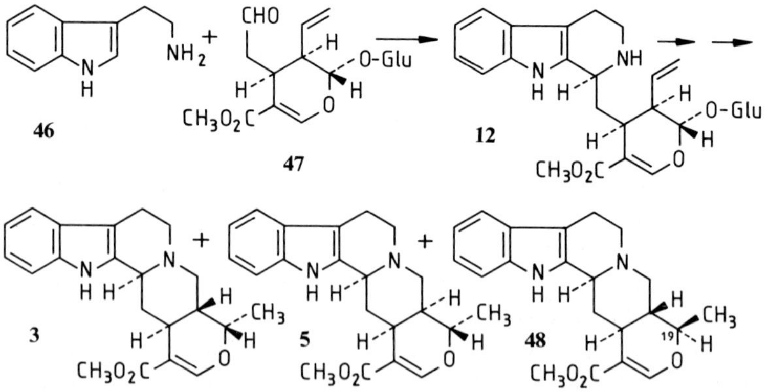

Scheme 1. Biogenetic formation of ajmalicine, tetrahydroalstonine and 19-epi-ajmalicine from tryptamine and secologanin

Although the later biogenetic stages in the production of the monomeric alkaloids are only partly known (*38–40*), it seems highly probable from the sequential appearance of alkaloidal types in the germinating plant (*41*) that the order of the biosynthesis is corynanthe, corynanthe-strychnos, aspidosperma, iboga. Further work is clearly needed in this difficult area.

Since the formation of the dimeric alkaloids vinblastine (**1**) and/or vincristine (**2**) has not been firmly demonstrated in cell suspension cultures (*vide supra*), we will not discuss their biogenetic pathway here. It will be referred to in part in section 2.2.8. which deals with the cell free system.

2.2. General Methods of *Catharanthus roseus* Cell Suspension Culture Work

2.2.1. Analytical Methods

The biotechnological application of cell cultures depends upon the development of stable cell lines yielding comparable or higher levels of the target alkaloids than the differentiated plant. The development of these cell lines, whether by the establishment of cell cultures from highly productive plants, the development of optimum cell culture conditions or the selection of strains producing high levels of the desired alkaloids, requires a proper analytical method – one which is not only sensitive and highly specific for the compound under investigation but permits the rapid analysis of a large number of samples.

Formerly, thin layer chromatography (*42*) was used for qualitative and quantitative determination of alkaloids in *C. roseus* plants and cell cultures but this method is insensitive, relatively slow, and requires the prepurification of samples. High pressure liquid chromatography and gas chromatography are more sensitive but also require prepurification of samples. The more recent application of radioimmunoassay (RIA) and enzyme-linked immunosorbent assay (ELISA) techniques to indole alkaloid determination has made possible the quantitation of these alkaloids with high specificity in the p-mol range (see e.g. refs. *16, 26, 43*). By allowing the analysis of several hundred nonpurified plant extracts in a single day, these methods are the only viable alternative in large-scale screening programmes. Nevertheless, the chromatographic methods (TLC, HPLC, GC), in conjunction with new, very effective sample clean-up (*44*) and preparation (*45*), continue to be more widely used than the more expensive RIA and ELISA methods.

2.2.2. Development of High Yielding Cell Lines

Before cell suspension cultures can be exploited for the industrial production of valuable substances, long and tedious laboratory work is required in order to develop cell lines with demonstrated high productivity and sufficient stability for large-scale fermentation.

The first stage in cell culture work is the establishment of high yielding cell lines through the selection of good source material. Proper selection, in turn, depends upon a systematic comparison of the contents of the secondary products in plants and in the cell cultures derived from them. In order to make systematic comparisons it is essential that the conditions of the experiments be identical.

In the case of *C. roseus* high producing cell lines could be developed based on the selection of high yielding differentiated plants (*16, 46*) as well as on the selection of plant varieties regardless of their biosynthetic capacity (*47*).

The culture process begins with the formation of callus from an explant of the source material. Though in some cases the callus can be derived from any part of the source plant without producing significant differences in metabolite content (*48*), high yielding cell lines of *C. roseus* have generally been initiated from calli derived from seeds, anther and leaf.

Cell lines are notoriously subject to variation, with effects on the alkaloid pattern as well as alkaloid levels (*17*); thus the establishment of high yielding and substrate-specific cell lines critically depends on the proper selection of variant lines. Because cells are highly diverse and even small aggregates within the culture are inhomogeneous, the effective selection method needs to be quick and highly sensitive to allow the rapid analysis of large numbers of small samples. With an insensitive and slow method (e.g. TLC), only an average value can be obtained for the biosynthetical capacity of fast growing, relatively large, inhomogeneous colonies, and the probability of an effective selection is low. The RIA and ELISA screening procedures fulfil the requirements of both speed and sensitivity and are entirely adequate for the establishment of productive cell lines (*26, 49*).

Fluorescence assay has proved a very effective selection method for fluorescence active compounds such as ajmalicine (**3**) and serpentine (**4**) (*16*), though it cannot give information about the ratio of the compounds. For final selection of a cell line this method needs to be followed by an exact analytical determination of the alkaloid content and ratio.

Once selected, high yielding lines have to be maintained for later experiments and large-scale fermentation. Unfortunately, despite reports about stable cell strains (*17*), the alkaloid pattern and the biosynthetic capacity tend to change over time. No satisfactory explanation of this phenomenon has yet been given; perhaps chromosomal aberration (*50*) is responsible. Cell strains serially subcultured under identical conditions sometimes show a certain degree of stability, but after the almost unavoidable changes in culture conditions (as in scale-up experiments for example) they tend to lose their stability. With repeated selection (*50*) the biosynthetic capacity can be retained but this is not a final solution to the problem. Cryopreservation (*51, 52*) might be helpful, but the variation and the loss in cell viability caused by the freezing technique itself would have to be overcome before the successful use of this method. Since industrial application of the cell suspension culture technique is impossible without the development and maintenance of stable strains, the solution of this problem necessarily must precede the scale-up experiments (*53*).

The final step in the evaluation of cell cultures is the isolation and quantification of the products. Plant secondary products are generally stored intracellularly, mostly in the vacuoles (*46, 54*) and are released into the culture medium in insignificant quantities. At present the most widely used isolation method for alkaloids is extraction from the filtered and freeze-dried cells, followed by purification and separation by chromatographic methods (e.g. TLC, HPLC, column chromatography). However, because the cells grow slowly it would be advantageous for an eventual industrial application if the release of intracellular products into the medium could be effected without destruction of the cells. Attempts have recently been made to elicit alkaloid release by application of pH gradients (*55*) and permeabilization of the cells (*56*), but more investigations are needed before either technique is ready for industrial use. After successful elicitation of the alkaloid release the products may be easily isolated by either known (e.g., extraction) or new (e.g. selective adsorption (*57*)) methods.

2.2.3. Growth and Alkaloid Production in Catharanthus roseus *Cell Suspension Cultures*

Besides the selection of high producing cell lines the development of favourable culture conditions is critical to the growth and biosynthetic capacity of cell cultures. It is well documented that cell cultures tend to respond with great changes in their biological and biosynthetic behaviour even to slight changes in the culture conditions. Worldwide, much effort has been devoted to determining the optimal growth parameters and production conditions for *C. roseus* cultures, but despite some promising results the work is far from complete. Different results have sometimes been obtained under similar conditions, genotypes have behaved differently under identical conditions and, most importantly, the clinically valuable target alkaloids have not been firmly detected in *C. roseus* cell suspension cultures. The task is great and the solution of these problems requires not only a formidable number of further experiments, but the parameters in the experiments must be more carefully controlled and monitored (*58*) than previously in order to get more instructive results and ones that are reproducible in other laboratories.

Some of the details of the growth and alkaloid production are known and described in the literature. The time course characteristics of the cell cultures are periodic because for the biosynthetic capacity of the cell lines to be maintained they have to be serially subcultured. In a single cycle after inoculation there is first a short resting stage during which the mitotic index is nearly zero. This initial decline of the mitotic activity is probably caused by shock due to transfer of cells to different culture conditions (*59*). Two to three days later the cells begin to divide fast. In this period of exponential growth the fresh and dry weights increase sharply and the mitotic index reaches a maximum. Later the mitotic index drops to zero, the fresh and dry weights reach their maximum and then begin a slight decrease (*60, 61*) probably owing to cell lysis (*16, 61*). The growth of the cells is accompanied by a decrease in extracellular acidity (*16, 17*) and after the cell weight reaches a maximum the pH remains at a constant level (*17*).

The alkaloid level and spectrum also change over time. The biosynthetically more complex alkaloids appear and reach their maximum level later than those belonging to the earlier stage of the biogenetic pathway, but later in the cycle they begin to equilibrate (*60*). After reaching a maximum the alkaloid level slowly declines to zero due to metabolism (*17*).

The relationship between cell growth and the alkaloid yield is not yet clear because relatively few systematic experiments have been carried out in this area and some of the results cannot be fully interpreted at present. By way of illustration, in different reactors under otherwise similar conditions the growth and the alkaloid production can be parallel and uncoupled (*16*), and results may depend on the experimental procedure, too (*47, 62*). The situation is further complicated by the lack of standardized methods of measurement (of growth for example (*59*)), so that the experimental data obtained by different research groups are not always comparable.

References, pp. 108–115

2.2.4. Effects of Culture Conditions on Growth and Alkaloid Production

The production of alkaloids in cell suspension cultures is governed by environmental conditions as well as by the genotype of the material. The influence of genotype on the alkaloid production has been discussed in the earlier parts of this article. In the following we look at the importance of the growth regime (nature of medium, presence of precursors, light, temperature, etc.) for growth and alkaloid production. Some of the results, we would note, are contradictory, but this can be attributed to the different experimental conditions (variation of cells, different experimental procedure, non-standardized measurements, etc.).

2.2.4.1. Medium Composition

One of the major advantage of plant cell cultures is that the cells can be grown in simple synthetic media. In recent years, considerable effort has gone into the search for media that would optimize cell growth and production of secondary metabolites. The task is complex since the chemical composition of a medium optimal for cell growth is not always optimal for production of secondary metabolites. A common approach to solving this problem is to use two different, separately optimized media, though there are also attempts to realize one-stage systems. To allow the effects of medium components on growth to be seen against the effects on alkaloid production we discuss the two types of effects together.

Carbon

A carbon source is required both for growth of the cells and for alkaloid production. Though a wide variety of substrates can be used as carbon source, the highest rate of utilization can usually be obtained with sucrose and glucose. The sucrose is extracellularly hydrolysed (*63, 64*) to glucose and fructose, of which glucose is preferentially utilized by the cells (*61, 63*). The controlling effect of the carbon source can be seen by comparing the results of experiments carried out under sucrose-limited and sucrose-rich conditions. In sucrose-limited medium the growth rate and the cell number are less than in sucrose-rich medium and the cells accumulate no alkaloids or only very small amounts. These results suggest that under carbon-limited conditions the available substrate is used entirely for cell growth and that pathways for metabolite production are inoperative (*47*). In sucrose-rich medium the cell senescence is reduced, cell mortality is very low, and alkaloid accumulation is greatly stimulated (*65, 66*). Under such conditions, the tryptophan having reached a maximum declines sharply, supporting the hypothesis (*67*) that increased alkaloid synthesis in sucrose-rich media can be attributed to tryptophan decarboxylase activity. The sucrose level also influences some other factors relevant to alkaloid formation in the cells such as the utilization of other nutrients, osmotic effect, etc.

Nitrogen

Nitrogen has opposite effects on growth and alkaloid production since it promotes growth and inhibits alkaloid production (*66, 68*). The inhibiting effect has not always been observed, however (*69*). The effect of nitrogen on alkaloid production is not independent of the carbon source available to the cells, so during experiments on medium components not only the nitrogen level but also the C/N ratio needs to be taken into account. The most typical nitrogen sources are nitrate and ammonia in the form of inorganic salts, but others (e.g. some of the amino acids) can be applied as well.

Phosphorus

In the case of *C. roseus*, growth promoting phosphate in concentrations commonly used for cell culture media has an inhibitory effect on indole alkaloid synthesis (*69, 70*). Almost all the extracellular phosphate is taken up from the medium and accumulated by cells within 48 h after inoculation (*71*). Though phosphate (used together with sucrose) prolongs the growth, extends the lag and exponential phases and increases the cell number and dry wt. yield (*72*), alkaloid production occurs only when intracellular phosphate has been exhausted below the inhibitory level and the alteration of enzyme activities caused by phosphate (*71*) has become insignificant.

The commonly used phosphorus sources are sodium dihydrogen phosphate and potassium dihydrogen phosphate.

Growth Regulators

Plant growth regulators affect not only growth and differentiation of cultured cells but also the secondary metabolism. Effects vary greatly with the quality and quantity of the applied substance. Thus 2,4-dichlorophenoxyacetic acid (2,4-D) and naphthylacetic acid (NAA) greatly suppress alkaloid formation, whereas benzyladenine, in the absence of an auxin source, induces relatively high alkaloid but low cell yields; furthermore, indole-3-acetic acid (IAA) gives relatively high levels of

References, pp. 108–115

both biomass and alkaloids (*16*). Yields can be increased by using these regulators in combination (*16, 73*) and applying two-stage cell culture systems where the biomass is first obtained in a high level auxin medium and then cells are transferred to an alkaloid production medium containing a cytokinin (*16*). An alternative approach might be the gradual addition of auxins to the medium, sufficient for biomass production but low enough for high alkaloid production (*69*).

Bioregulators

Some synthetic bioregulators are stimulators of isoprenoid biosynthesis (*74*). As secologanin (**47**) is an important isoprenoid precursor of indole alkaloid formation, a number of bioregulators have been investigated for their effects on alkaloid production of *C. roseus* cell cultures (*75, 76*). The effects of the added compounds were different but some of them increased the alkaloid yield without significantly affecting parameters like rate of cell growth, cell biomass, initial pH, etc.

Medium pH

Despite the negative results for some cells lines (*59*), the alkaloid yield seems to be affected by the pH of the medium (*77, 78*). The mechanism of this effect is only partly understood but probably the alkaloid accumulation and storage system (*54, 55*) are influenced by the size of the pH gradient between the medium and vacuoles.

Apart from the above mentioned factors, there are other components of the medium such as vitamins, trace elements, etc. which play an important role in alkaloid production. However, many experiments remain to be done to develop an optimal medium composition. In the end, medium modification may turn out to be the key in producing those clinically and synthetically important alkaloids which to date have not been detected in *C. roseus* cell suspension cultures.

2.2.4.2. Addition of Precursors

An exogenous supply of a biosynthetic precursor to the culture medium may increase the yield of the final product where the productivity is limited by lack of the precursor. In the case of *C. roseus* the limiting factor in alkaloid accumulation has not been defined and may involve one or both of the indole and terpenoid pathways. The reports in the literature concerning the effects of precursors on alkaloid production are contradictory. Thus, adding tryptamine (**46**) or tryptophan or both to the medium has occasioned different responses in cells, in some cases stimulation of alkaloid production (*16, 46*) and in others repression *(79)* or no change *(80)*. The results were negative when cells were stimulated to overproduce these precursors. Though selected cell lines resistant to tryptophan analogues were found to accumulate much higher levels of tryptamine **(46)** and/or tryptophan, enhanced alkaloid accumulation was not observed *(67, 81, 82)*. Addition of precursors of the terpenoid pathway (e.g. loganin, secologanin **(47)**) in some cases increased alkaloid levels *(16, 80)*, but in other cases effects were insignificant *(83)* or inhibitory *(16)*.

Clearly no generalization of the positive or negative effects of any given precursors should be claimed unless the strains *(46)* and the culture parameters *(80)* are exactly defined and characterized.

2.2.4.3. Light

Light affects both growth and metabolite production of cell cultures though the effects largely depend on the medium composition and culture method. In the case of heterotrophic *C. roseus* cell cultures, the formation of anthocyanins *(68, 84)* and the indole alkaloids serpentine **(4)** *(68, 79)* and catharanthine **(38)** *(69)* was stimulated in light, while the level of ajmalicine **(3)** was higher than that of serpentine **(4)** *(66, 69)* in cultures grown in the dark. The extracellular alkaloid level is higher in the dark than under illumination *(69)*.

As cell suspension cultures require a more carefully defined growth regime than plants, a better understanding of the role of light in indole alkaloid biosynthesis would be helpful. Another useful approach might be investigations on photoautotrophic *C. roseus* cell cultures as recently initiated *(85)*.

2.2.4.4. Temperature

Growth and alkaloid production are differently affected by the culture temperature. Above 40–45° C cells die; no growth occurs at 10° C but cells remain viable. The growth rate is of a maximum at 35° C *(86)* and the rate is constant between 27° C and 35° C *(87)*. Biomass yields are maximal at about 25° C. Although 25° C is not the optimal temperature for alkaloid accumulation, this problem might be solved by small seasonal variations in the average culture temperature *(86)*.

Divergent results have been obtained regarding alkaloid accumulation at lower temperatures; both inhibiting *(86)* and stimulating *(87)* effects have been claimed. However, the experimental conditions were different and many other factors such as medium composition, biomass yield, cell number, capacity of the strain, etc. have to be taken into account for an exact comparison of the results.

References, pp. 108–115

2.2.4.5. Gaseous Environment

The gaseous environment, mostly the availability of oxygen and carbon dioxide, also plays an important role in the metabolic processes of plant cell cultures. Under conditions of unlimited oxygen a high level of carbon source utilization can be achieved, while oxygen limitation markedly reduces the conversion rate (*88*). Similar results can be obtained under carbon dioxide enriched and limited-conditions (*89, 90*). Above a critical level both oxygen and carbon dioxide suppress the growth.

2.2.5. One-stage Systems

Until recently, transfer of cells to alkaloid production medium has been the only effective method of inducing alkaloid accumulation and reproduction of enzymes involved in the biosynthesis. While this separation of growth and alkaloid production facilitates optimization of culture conditions, one-stage techniques with a shorter culture period offer a simpler and more cost-effective approach for eventual application (*73, 91*). Some success has been achieved in developing culture media that contain an appropriate balance of nutritional and hormonal components to allow sustained cell growth and high alkaloid accumulation. This system becomes even more effective when used in combination with elicitors (*92, 93*) or other compounds (*94, 95*) that increase the alkaloid production. Costs can be further reduced by applying the one-stage method in semi-continuous systems (*96*).

2.2.6. Large-scale Fermentation

Special reactors are required for cultivation of plant cell suspension cultures. The main problem here is the high viscosity combined with the sensitivity of cells to shear stress. With gentle agitation the gaseous transfer is very low, the culture becomes highly heterogeneous and the cells collect into relatively large aggregates and motionless regions. With too vigorous agitation the damage to cells becomes unacceptable and gas concentration may reach the inhibitory level. In *C. roseus* cell cultures both mechanically agitated (*53*) and air-lift (*97*) reactors are used. Each method has its advantages and disadvantages. A special problem connected with the air-lift reactor is the growth inhibiting effect of high aeration rates (*90, 98, 99*). An alternative approach to the mixing problem might be modification of the medium in such a way that aggregates remain smaller even while cells sustain their biosynthetic capacity (*61*).

Yet a further problem is the sensitivity of plant cells to infection, something that is more difficult to tackle in large-scale fermentation than under laboratory conditions. Whatever the technical difficulties, both the inoculum and the reactor have to be carefully sterilized in order to avoid failure due to infection (*53*).

But despite the above and other technical difficulties, the main problem remains the lack of high yielding stable cell lines, and the solution of this problem would appear to be more difficult than optimization of the scale-up or the development of continuous cell culture systems (*100, 101*).

2.2.7. Immobilized Cell Systems

Because cells grow slowly it would be advantageous to be able to utilize the biomass for extended periods. One promising approach is the application of immobilized cell systems, which have the advantage over free cell systems of a prolonged operational life. In combination with elicitation of alkaloid release (*56, 102*), continuous processes could be developed for industrial application. Moreover, by feeding of precursors it might be possible to produce metabolites not normally synthesized by the cells. A number of methods for cell immobilization are known. In the case of *C. roseus*, cells are commonly entrapped in a gel matrix, usually calcium alginate. Cells of *C. roseus* entrapped in this way show stable viability and biosynthetic capacity over long periods (*103, 104*) and *de novo* synthesis as well as synthesis from precursors (*105–106*) very possibly could be realized with this culture technique.

2.2.8. Cell Free Systems

Studies on the biogenetic pathway of indole alkaloid biosynthesis have turned the attention of many researchers to cell cultures, where the enzymes of the biosynthesis are present at much higher levels than in the intact plant. High incorporation of precursors have been realized under controlled conditions. A further step has been the development of cell free systems in which enzymatic reactions can be studied without the disturbing effects of cell metabolic processes. With the help of this method a number of late steps in the pathway to the dimeric alkaloids have been clarified.

The first important of the late steps in this pathway is the coupling reaction of the two monomeric units vindoline (**37**) and catharanthine

(38). Incubation of (37) and (38) with cell free extract of *C. roseus* results in 3′,4′-anhydrovinblastine (49) (*107*) (Scheme 2); and (49) has been shown in cell free experiments to be an intermediate in the biosynthesis of vinblastine-type alkaloids (*107–118*).

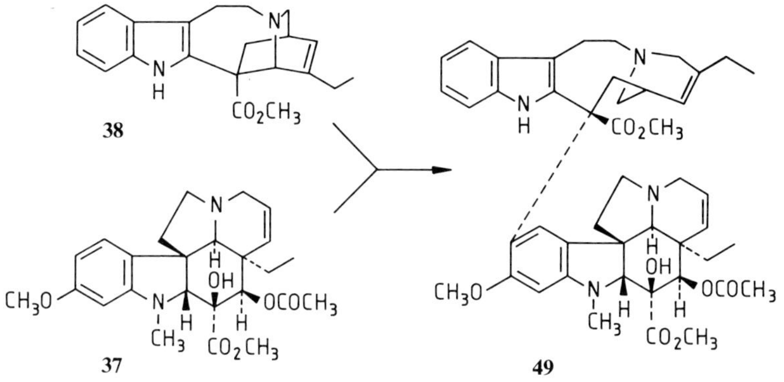

Scheme 2. Coupling of catharanthine and vindoline to 3′,4′-anhydro-vinblastine

The fact that dimeric alkaloids can be formed in cell free extracts of *C. roseus* cell suspension cultures suggests that these clinically valuable compounds could be produced from cell cultures but that probably one or more of the earlier steps of the biosynthesis are blocked.

Cell free systems also provide a potentially important medium for the biosynthesis of the alkaloids on the pathway. Vindoline (37) formation was claimed (*119*) after incubation of tryptamine (46) and secologanin (47) with cell free extract of *C. roseus*. The results were later contested (*120*).

The complexity of the crude cell free extracts and the instability of the enzyme activities promise difficulties for any commercial biotechnological application of cell free systems. In an attempt to overcome such difficulties a number of enzymes have been purified (*36, 38, 121*) and high enzyme stability has been achieved by application of immobilization technique (*122, 123*).

3. Conclusions

Recent years have seen great effort invested in the possibility of producing the pharmacologically important dimeric alkaloids and their biosynthetic precursors by cell culture techniques. Despite the promising results, a number of important difficulties have not yet been solved.

One of the greatest disappointments in the cell culture work is the entire lack of or formation of only very small amounts of the main target alkaloids in cultures thus far (*vide supra*). There are some important biological and technical problems to be overcome as well. Nevertheless, our understanding of the controlling factors involved in the secondary metabolism has increased markedly and there are real grounds for optimism that the cell culture technique can one day be applied on an industrial scale for production of pharmacologically valuable alkaloids of *C. roseus*.

References

1. SVOBODA, G.H., and D.A. BLAKE: The Phytochemistry and Pharmacology of *Catharanthus roseus* (L.) G. Don. In: The *Catharanthus* Alkaloids (TAYLOR, W.I., and N.R. FARNSWORTH, Eds.), p. 45. New York: Marcel Dekker, 1975.
2. DECONTI, R.C., and W.A. CREASEY: Clinical Aspects of the Dimeric *Catharanthus* Alkaloids. In: The *Catharanthus* Alkaloids (TAYLOR, W.I., and N.R. FARNSWORTH, Eds.), p. 237. New York: Marcel Dekker. 1975.
3. CORDELL, G.A.: The Botanical, Chemical, Biosynthetic and Pharmacologic Aspects of *Catharanthus roseus* (L.) G. Don (Apocynaceae). In: Recent Advances in Natural Product Research (Woo, W.S., and B.H. HAN, Eds.), p. 65. Seoul: Seoul National University Press. 1980.
4. LOUNASMAA, M., and A. NEMES: The Synthesis of Bis-indole Alkaloids and their Derivatives. Tetrahedron **38**, 223 (1982).
5. POTIER, P.: Synthesis of Bio-active Substances: Recent Examples. In: Stereoselective Synthesis of Natural Products (BARTMANN, W., and E. WINTERFELDT, Eds.) p. 19. Amsterdam: Excerpta Medica. 1978.
6. – Synthesis of the Antitumor Dimeric Indole Alkaloids from *Catharanthus* Species (Vinblastine Group). J. Nat. Prod. **43**, 72 (1980).
7. CORDELL, G.A., J.E. SAXTON: Bisindole Alkaloids. In: The Alkaloids (RODRIGO, R.G.A. Ed.), Vol. 20, p. 1. New York: Academic Press. 1981.
8. – The Bisindole Alkaloids. In: Indoles. The Monoterpenoid Indole Alkaloids (SAXTON, J.E. Ed.), p. 539. New York: John Wiley. 1983.
9. RAUCHER, S., B.L. BRAY, and R.F. LAWRENCE: Synthesis of ($\pm$)-Catharanthine, ($+$)-Anhydrovinblastine, and ($-$)-Anhydrovincovaline. J. Am. Chem. Soc. **109**, 442 (1987).
10. VUKOVIC, J., A.E. GOODBODY, J.P. KUTNEY, and M. MISAWA: Production of 3',4'-Anhydrovinblastine: A Unique Chemical Synthesis. Tetrahedron **44**, 325 (1988).
11. KUTNEY, J.P., L.S.L. CHOI, J. NAKANO, H. TSUKAMOTO, M. MCHUGH, and C.A. BOULET: A Highly Efficient and Commercially Important Synthesis of the Antitumor *Catharanthus* Alkaloids Vinblastine and Leurosidine from Catharanthine and Vindoline. Heterocycles **27**, 1845 (1988).
12. MIURA, Y., and K. HIRATA: An Organ Culture of *Catharanthus roseus* Capable of Producing Substantial Amount of Indole Alkaloids. Eur. Pat. Appl. EP 0 200 225 A2 (1986).
13. MIURA, Y., K. HIRATA, and N. KURANO: Isolation of Vinblastine in Callus Culture with Differentiated Roots of *Catharanthus roseus* (L.) G. Don. Agric. Biol. Chem. **51**, 611 (1987).
14. LE MEN, J., and W.I. TAYLOR: A Uniform Numbering System for Indole Alkaloids. Experientia **21**, 508 (1965).

15. HARRIS, A.L. H. NYLUND, and D.P. CAREW: Tissue Culture Studies of Certain Members of the Apocynaceae. J. Nat. Prod. (Lloydia) **27**, 322 (1964).
16. ZENK, M.H., H. EL-SHAGI, H. ARENS, J. STÖCKIGT, E. WEILER, and B. DEUS: Formation of the Indole Alkaloids Serpentine and Ajmalicine in Cell Suspension Cultures of *Catharanthus roseus*. In: Plant Tissue Culture and Its Bio-Technological Application (BARZ, W., E. REINHARD, and M.H. ZENK, Eds.), p. 27. Berlin-Heidelberg-New York: Springer. 1977.
17. KURZ, W.G.W., K.B. CHATSON, F. CONSTABEL, J.P. KUTNEY, L.S.L. CHOI, P. KOLODZIEJCZYK, S.K. SLEIGH, K.L. STUART, and B.R. WORTH: Alkaloid Production in *Catharanthus roseus* Cell Cultures: Initial Studies on Cell Lines and their Alkaloid Content. Phytochemistry **19**, 2583 (1980).
18. KUTNEY, J.P., L.S.L. CHOI, P. KOLODZIEJCZYK, S.K. SLEIGH, K.L. STUART, B.R. WORTH, W.G.W. KURZ, K.B. CHATSON, and F. CONSTABEL: Alkaloid Production in *Catharanthus roseus* Cell Cultures. V. Alkaloids from the 176G, 299Y, 340Y and 951G Cell Lines. J. Nat. Prod. **44**, 536 (1981).
19. KURZ, W.G.W., K.B. CHATSON, F. CONSTABEL, J.P. KUTNEY, L.S.L. CHOI, P. KOLODZIEJCZYK, S.K. SLEIGH, and K.L. STUART: The Production of Catharanthine and Other Indole Alkaloids by Cell Suspension Cultures of *Catharanthus roseus*. Planta Med. **39**, 284 (1980).
20. STÖCKIGT, J., and H.J. SOLL: Indole Alkaloids from Cell Suspension Cultures of *Catharanthus roseus* and *C. ovalis*. Planta Med. **40**, 22 (1980) and references therein.
21. KOHL, W., B. WITTE, and G. HÖFLE: Alkaloide aus *Catharanthus roseus*-Zellkulturen II. Z. Naturforsch. **36b**, 1153 (1981).
22. – – – : Alkaloide aus *Catharanthus roseus*-Zellkulturen III. Z. Naturforsch. **37b**, 1346 (1982).
23. KOHL, W., B. WITTE, W.S. SHELDRICK, and G. HÖFLE: Indolalkaloide aus *Catharanthus roseus*-Zellkulturen IV. 16*R*-19,20-*E*-Isositsirikin, 16*R*-19,20-*Z*-Isositsirikin und 21-Hydroxycyclolochnerin. Planta Med. **50**, 242 (1984).
24. GUERITTE, F., N. LANGLOIS, and V. PETIARD: Métabolites Secondaires Isolés d'Une Culture de Tissus de *Catharanthus roseus*. J. Nat. Prod. **46**, 144 (1983).
25. PETIARD, V., and D. COURTOIS: Recent Advances in Research for Novel Alkaloids in Apocynaceae Tissue Cultures. Physiol. Vég. **21**, 217 (1983).
26. LAPINJOKI, S., H. VERÄJÄNKORVA, J. HEISKANEN, M. NISKANEN, A. HUHTIKANGAS, and M. LOUNASMAA: Immunoanalytical Methods for Screening Vindoline from *Catharanthus roseus* Cell Cultures. Planta Med. **53**, 565 (1987).
27. NAARANLAHTI, T., S.P. LAPINJOKI, A. HUHTIKANGAS, L. TOIVONEN, U. KURTEN, V. KAUPPINEN, and M. LOUNASMAA: Mass Spectral Evidence on the Existence of Vindoline in Heterothropic *Catharanthus roseus* Cell Cultures. Planta Med. **55**, 155 (1989).
28. SCOTT, A.I., H. MIZUKAMI, T. HIRATA, and S.L. LEE: Formation of Catharanthine, Akuammicine and Vindoline in *Catharanthus roseus* Suspension Cells. Phytochemistry **19**, 488 (1980).
29. GRÖGER, D.: Alkaloids Derived from Tryptophan. In: Biochemistry of Alkaloids (MOTHES, K., H.R. SCHUTTE, and M. LUCKNER, Eds.), p. 272. Weinheim: VCH Verlagsgesellschaft, 1985.
30. ATTA-UR-RAHMAN and A. BASHA: Biosynthesis of Indole Alkaloids. Oxford: Clarendon Press. 1983.
31. STÖCKIGT, J., and M.H. ZENK: Strictosidine (Isovincoside): the Key Intermediate in the Biosynthesis of Monoterpenoid Indole Alkaloids. J.C.S. Chem. Commun. **1977**, 646.
32. TREIMER, J.F., and M.H. ZENK: Enzymic Synthesis of Corynanthe-type Alkaloids in Cell Cultures of *Catharanthus roseus*: Quantitation by Radioimmunoassay. Phytochemistry **17**, 227 (1978).

33. Scott, A.I., S.L. Lee, M. G., Culver, W. Wan, T. Hirata, F. Guerrite, R.L. Baxter, H. Nordlöv, C.A. Dorschel, H. Mizukami, and N.E. Mackenzie: Indole Alkaloid Biosynthesis. Heterocycles **15**, 1257 (1981).
34. Scott, A.I., S.L. Lee, P. De Capite, M.G. Culver, and C.R. Hutchinson: The Role of Isovincoside (Strictosidine) in the Biosynthesis of the Indole Alkaloids. Heterocycles **7**, 979 (1977).
35. Zenk, M.H.: Enzymatic Synthesis of Ajmalicine and Related Indole Alkaloids. J. Nat. Prod. **43**, 438 (1980).
36. Treimer, J.F., and M.H. Zenk: Strictosidine Synthase from Cell Cultures of Apocynaceae Plants. FEBS Lett. **97**, 159 (1979).
37. Mizukami, H., H. Nordlöv, S.L. Lee, and A.I. Scott: Purification and Properties of Strictosidine Synthetase (an Enzyme Condensing Tryptamine and Secologanin) from *Catharanthus roseus* Cultured Cells. Biochemistry **18**, 3760 (1979).
38. Deluca, V., J. Balsevich, R.T. Tyler, and W.G.W. Kurz: Characterization of a Novel N-Methyltransferase (NMT) from *Catharanthus roseus* Plants. Plant Cell Reports **6**, 458 (1987).
39. Fahn, W., H. Gundlach, B. Deus-Neumann, and J. Stöckigt: Late Enzyme of Vindoline Biosynthesis. Acetyl-CoA: 17-*O*-Deacetylvindoline 17-*O*-Acetal-transferase. Plant Cell Reports **4**, 333 (1985).
40. Fahn, W., E. Laussermair, B. Deus-Neumann, and J. Stöckigt: Late Enzyme of Vindoline Biosynthesis. *S*-Adenosyl-*L*-methionine: 11-*O*-Demethyl-17-*O*-deacetylvindoline 11-*O*-Methyltransferase and Unspecific Acetylesterase. Plant Cell Reports **4**, 337 (1985).
41. Scott, A.I., P.C. Cherry, and A.A. Qureshi: Mechanisms of Indole Alkaloid Biosynthesis. The *Corynanthe-Strychnos* Relationship. J. Am. Chem. Soc. **91**, 4932 (1969).
42. Farnsworth, N.R., R.N. Blomster, D. Damratoski, W.A. Meer, and L.V. Cammarato: *Catharanthus* Alkaloids. VI. Evaluation by Means of Thin-layer Chromatography and Ceric Ammonium Sulfate Spray Reagent. J. Nat. Prod. (Lloydia) **27**, 302 (1964).
43. Deus-Neumann, B., J. Stöckigt, and M.H. Zenk: Radioimmunoassay for the Quantitative Determination of Catharanthine. Planta Med. **53**, 184 (1987).
44. Kohl, W., B. Witte, and G. Höfle: Quantitative und Qualitative HPLC-Analytik von Indolalkaloiden aus *Catharanthus roseus*-Zellkulturen. Planta Med. **47**, 177 (1983).
45. Kutney, J.P., L.S.L. Choi, P. Kolodziejczyk, S.K. Sleigh, K.L. Stuart, B.R. Worth, W.G.W. Kurz, K.B. Chatson, and F. Constabel: Alkaloid Production in *Catharanthus roseus* Cell Cultures: Isolation and Characterization of Alkaloids from One Cell Line. Phytochemistry **19**, 2589 (1980).
46. Deus, B., and M.H. Zenk: Exploitation of Plant Cells for the Production of Natural Compounds. Biotechnol. Bioeng. **24**, 1965 (1982).
47. Kurz, W.G.W., and F. Constabel: Aspects Affecting Biosynthesis and Biotransformation of Secondary Metabolites in Plant Cell Cultures. CRC Critical Revievs in Biotechnology **2**, 105 (1985).
48. Sasse, F., U. Heckenberg, and J. Berlin: Accumulation of β-Carboline Alkaloids and Serotonin by Cell Cultures of *Peganum harmala* L.I. Correlation between Plants and Cell Cultures and Influence of Medium Constituents. Plant Physiol. **69**, 400 (1982).
49. Huhtikangas, A., T.Lehtola, S. Lapinjoki, and M. Lounasmaa: Specific Radioimmunoassay for Vincristine. Planta Med. 53, 85 (1987), and references therein.
50. Deus-Neumann, B., and M.H. Zenk: Instability of Indole Alkaloid Production in *Catharanthus roseus* Cell Suspension Cultures. Planta Med. **50**, 427 (1984).

51. CHEN, T.H.H., K.K. KARTHA, N.L. LEUNG, W.G.W. KURZ, K.B. CHATSON, and F. CONSTABEL: Freezing Characteristics of Cultured *Catharanthus roseus* (L.) G. Don Cells Treated with Dimethylsulfoxide and Sorbitol in Relation to Cryopreservation. Plant Physiol **75**, 720 (1984).
52. – – – – – – Cryopreservation of Alkaloid-producing Cell Cultures of Periwinkle (*Catharanthus roseus*). Plant Physiol. **75**, 726 (1984).
53. SCHIEL, O., and J. BERLIN: Large Scale Fermentation and Alkaloid Production of Cell Suspension Cultures of *Catharanthus roseus*. Plant Cell Tissue Org. Cult. **8**, 153 (1987).
54. NEUMANN, D., G. KRAUSS, M. HIEKE, and D. GRÖGER: Indole Alkaloid Formation and Storage in Cell Suspension Cultures of *Catharanthus roseus*. Planta Med. **48**, 20 (1983).
55. BOUYSSOU, H., A. PAREILLEUX, and G. MARIGO: The Role of pH Gradients Across the Plasmalemma of *Catharanthus roseus* and Its Involvement in the Release of Alkaloids. Plant Cell Tissue Org. Cult. **10**, 91 (1987).
56. BRODELIUS, P., and K. NILSSON: Permeabilization of Immobilized Plant Cells, Resulting in Release of Intracellularly Stored Products with Preserved Cell Viability. Eur. J. Appl. Microbiol. Biotechnol. **17**, 275 (1983).
57. PAYNE, G.F., and M.L. SHULER: Alkaloid Recovery for Plant Cell Systems. Biotechnol. Bioeng. Symp. **15**, 633 (1986).
58. KUTNEY, J.P.: Studies in Plant Culture. The Synthesis and Biosynthesis of Indole Alkaloids. Heterocycles **25**, 617 (1987).
59. KURZ, W.G.W., K.B. CHATSON, F. CONSTABEL, J.P. KUTNEY, L.S.L. CHOI, P. KOLODZIEJCZYK, S.K. SLEIGH, K.L. STUART, and B.R. WORTH: Alkaloid Production in *Catharanthus roseus* Cell Cultures VIII. Characterisation of the PRL/200 Cell Line. Planta Med. **42**, 22 (1981).
60. – – – – – – – – Alkaloid Production in *Catharanthus roseus* Cell Cultures. IV. Characterization of the 953 Cell Line. Helv. Chim. Acta **63**, 1891 (1980).
61. DRAPEAU, D., H.W. BLANCH, and C.R. WILKE: Growth Kinetics of *Dioscorea deltoidea* and *Catharanthus roseus* in Batch Culture. Biotechnol. Bioeng. **28**, 1555 (1986).
62. STAFFORD, A., L. SMITH, and M.W. FOWLER: Regulation of Product Synthesis in Cell Cultures of *Catharanthus roseus* (L.) G. Don. Plant Cell Tissue Org. Cult. **4**, 83 (1985).
63. MORRIS, P., and M.W. FOWLER: Sucrose Utilization by Cell Suspension Cultures of *Catharanthus roseus* G. Don. Biochem. Soc. Trans. **8**, 630 (1980).
64. STAFFORD, A., and M.W. FOWLER: Effect of Carbon and Nitrogen Growth Limitation upon Nutrient Uptake and Metabolism in Batch Cultures of *Catharanthus roseus* (L.) G. Don. Plant Cell Tissue Org. Cult. **2**, 239 (1983).
65. MERILLON, J.M., M. RIDEAU, and J.C. CHENIEUX: Influence of Sucrose on Levels of Ajmalicine, Serpentine, and Tryptamine in *Catharanthus roseus* Cells *in vitro*. Planta Med. **50**, 497 (1984).
66. KNOBLOCH, K.H., and J. BERLIN: Influence of Medium Composition on the Formation of Secondary Compounds in Cell Suspension Cultures of *Catharanthus roseus* (L.). G. Don. Z. Naturforsch. **35c**, 551 (1980).
67. SCHALLENBERG, J., and J. BERLIN: 5-Methyltryptophan Resistant Cells of *Catharanthus roseus*. Z. Naturforsch. **34c**, 541 (1979).
68. KNOBLOCH, K.H., G. BAST, and J. BERLIN: Medium- and Light-induced Formation of Serpentine and Anthocyanins in Cell Suspension Cultures of *Catharanthus roseus*. Phytochemistry **21**, 591 (1982).
69. DRAPEAU, D., H.W. BLANCH, and C.R. WILKE: Ajmalicine, Serpentine, and Catharanthine Accumulation in *Catharanthus roseus* Bioreactor Cultures. Planta Med. **53**, 373 (1987).

70. KNOBLOCH, K.H., and J. BERLIN: Effects of Media Constituents on the Formation of Secondary Products in Cell Suspension Cultures of *Catharanthus roseus*. In: Advances in Biotechnology (MOO-YOUNG, M., and C.W. ROBINSON, Eds.), Vol. 1, p. 129. Toronto, Oxford, New York, Sydney, Paris, Frankfurt: Pergamon Press. 1981.
71. – – Influence of Phosphate on the Formation of the Indole Alkaloids and Phenolic Compounds in Cell Suspension Cultures of *Catharanthus roseus*. I. Comparison on Enzyme Activities and Product Accumulation. Plant Cell Tissue Org. Cult. **2**, 333 (1983).
72. MACCARTHY, J.J., and D. RATCLIFFE: The Effect of Nutrient Medium Composition on the Growth Cycle of *Catharanthus roseus* G. Don Cells Grown in Batch Culture. J. Exp. Bot. **31**, 1315 (1980).
73. MORRIS, P.: Regulation of Product Sythesis in Cell Cultures of *Catharanthus roseus*. II: Comparison of Production Media. Planta Med. **52**, 121 (1986).
74. YOKOYAMA, H., E.P. HAYMAN, W.J. HSU, and S.M. POLING: Chemical Bioinduction of Rubber in Guayule Plant. Science **197**, 1076 (1977).
75. LEE, S.L., K.D. CHENG, and A.I. SCOTT: Effects of Bioregulators on Indole Alkaloid Biosynthesis in *Catharanthus roseus* Cell Culture. Phytochemistry **20**, 1841 (1981).
76. KUTNEY, J.P., B. AWERYN, K.B. CHATSON, L.S.L. CHOI, and W.G.W. KURZ: Alkaloid Production in *Catharanthus roseus* (L.). G. Don Cell Cultures. XIII. Effects of Bioregulators on Indole Alkaloid Biosynthesis. Plant Cell Reports **4**, 259 (1985).
77. KUTNEY, J.P., L.S.L. CHOI, P. KOLODZIEJCZYK, S.K. SLEIGH, K.L. STUART, B.R. WORTH, W.G.W. KURZ, K.B. CHATSON, and F. CONSTABEL: Alkaloid Production in *Catharanthus roseus* Cell Cultures. VII. Effect of Parameter Changes and Catabolism Studies on Cell Line PRL No 953. Helv. Chim. Acta **64**, 1837 (1981).
78. DÖLLER, G.: Influence of the Medium on the Production of Serpentine by Suspension Cultures of *Catharanthus roseus* (L.) G. Don. In: Production of Natural Compounds by Cell Culture Methods (ALFERMANN, A.W., and E. REINHARD, Eds.), p. 109. Munich: G.f.S.U. 1978.
79. DÖLLER, G, A.W. ALFERMANN, and E. REINHARD: Production of Indole Alkaloids in Tissue Cultures of *Catharanthus roseus*. Planta Med. **30**, 14 (1976).
80. MERILLON, J.M., P. DOIREAU, A. GUILLOT, J.C. CHENIEUX, and M.RIDEAU: Indole Alkaloid Accumulation and Tryptophan Decarboxylase Activity in *Catharanthus roseus* Cells Cultured in Three Different Media. Plant Cell Reports **5**, 23 (1986).
81. SCOTT, A.I., H. MIZUKAMI, and S.L. LEE: Characterization of a 5-Methyltryptophan Resistant Strain of *Catharanthus roseus* Cultured Cells. Phytochemistry **18**, 795 (1979).
82. SASSE, F., M. BUCHHOLZ, and J. BERLIN: Selection of Cell Lines of *Catharanthus roseus* with Increased Tryptophan Decarboxylase Activity. Z. Naturforsch. **38c**, 916 (1983).
83. KRUEGER, R.J., and D.P. CAREW: *Catharanthus roseus* Tissue Culture: The Effects of Precursors on Growth and Alkaloid Production. J. Nat. Prod. (Lloydia) **41**, 327 (1978).
84. CAREW, D.P., R.J. KRUEGER: Anthocyanidins of *Catharanthus roseus* Callus Cultures. Phytochemistry **15**, 442 (1976).
85. TYLER, R.T., W.G.W. KURZ, and B.D. PANCHUK: Photoautotrophic Cell Suspension Cultures of Periwinkle (*Catharanthus roseus* (L.) G. Don): Transition from Heterotrophic to Photoautotrophic Growth. Plant Cell Reports **3**, 195 (1986).
86. MORRIS, P.: Regulation of Product Synthesis in Cell Cultures of *Catharanthus roseus*. Effect of Culture Temperature. Plant Cell Reports **5**, 427 (1986).
87. COURTOIS, D., and J. GUERN: Temperature Response of *Catharanthus roseus* Cells Cultivated in Liquid Medium. Plant Science Lett. **17**, 473 (1980).

88. PAREILLEUX, A., and R. VINAS: Influence of the Aeration Rate on the Growth Yield in SUSPENSION Cultures of *Catharanthus roseus* (L.) G. Don. J. Ferment. Technol. **61**, 429 (1983).
89. MAUREL, B., and A. PAREILLEUX: Effect of Carbon Dioxide on the Growth of Cell Suspension of *Catharanthus roseus*. Biotechnol. Lett. **7**, 313 (1985).
90. DUCOS, J.P., and A. PAREILLEUX: Effect of Aeration Rate and Influence of pCO_2 in Large-scale Cultures of *Catharanthus roseus* Cells. Appl. Microbiol. Biotechnol. **25**, 101 (1986).
91. SMITH, J.I., A.A. QUESNEL, N.J. SMART, M. MISAWA, and W.G.W. KURZ: The Development of a Single-stage Growth and Indole Alkaloid Production Medium for *Catharanthus roseus* (L.). G. Don Suspension Cultures. Enzyme Microb. Technol. **9**, 466 (1987).
92. EILERT, U., V. DELUCA, F. CONSTABEL, and W.G.W. KURZ: Elicitor-mediated Induction of Tryptophan Decarboxylase and Strictosidine Synthase Activities in Cell Suspension Cultures of *Catharanthus roseus*. Arch. Biochem. Biophys. **254**, 491 (1987).
93. EILERT, U., F. CONSTABEL, and W.G.W. KURZ: Elicitor-stimulation of Monoterpene Indole Alkaloid Formation in Suspension Cultures of *Catharanthus roseus*. J. Plant Physiol. **126**, 11 (1986).
94. SMITH, J.I., N.J. SMART, M. MISAWA, W.G.W. KURZ, S.G. TALLEVI, and F. DICOSMO: Increased Accumulation of Indole Alkaloids by Some Cell Lines of *Catharanthus roseus* in Response to Addition of Vanadyl Sulphate. Plant Cell Reports **6**, 142 (1987).
95. SMITH, J.I., N.J. SMART, W.G.W. KURZ, and M. MISAWA: The Use of Organic and Inorganic Compounds to Increase the Accumulation of Indole Alkaloids in *Catharanthus roseus* (L.) G. Don Cell Suspension Cultures. J. Exp. Bot. **38**, 1501 (1987).
96. CONSTABEL, F., W.G.W. KURZ, and U. EILERT: Semi-continuous Production and Secretion of Phytochemicals by Plant Cell Culture with Successive Elicitation. Eur. Pat. Appl. EP 226, 354.
97. SMART, N.J., and M.W. FOWLER: Mass Cultivation of *Catharanthus roseus* Cells Using a Nonmechanically Agitated Bioreactor. Appl. Biochem. Biotechnol. **9**, 209 (1984).
98. HEGARTY, P.K., N.J. SMART, A.H. SCRAGG, and M.W. FOWLER: The Aeration of *Catharanthus roseus* L. G. Don Suspension Cultures in Airlift Bioreactors: the Inhibitory Effect at High Aeration Rates on Culture Growth. J. Exp. Bot. **37**, 1911 (1986).
99. SMART, N.J., and M.W. FOWLER: Effect of Aeration on Large-scale Cultures of Plant Cells. Biotechnol. Lett. **3**, 171 (1981).
100. PAREILLEUX, A., and R. VINAS: A Study on the Alkaloid Production by Resting Cell Suspension of *Catharanthus roseus* in a Continuous Flow Reactor. Appl. Microbiol. Biotechnol. **19**, 316 (1984).
101. ROSEVEAR, A., and S.D. ROE: Secondary Metabolite Manufacture Using Immobilized Cells and Affinity Chromatography in a Continuous Flow Process. Ger. Offen. De 3, 616, 357.
102. FELIX, H., P. BRODELIUS, and K. MOSBACH: Enzyme Activities of the Primary and Secondary Metabolism of Simultaneously Permeabilized and Immobilized Plant Cells. Anal. Biochem. **116**, 462 (1981).
103. MAJERUS, F., and A. PAREILLEUX: Alkaloid Accumulation in Ca-alginate Entrapped Cells of *Catharanthus roseus*: Using Limiting Growth Medium. Plant Cell Reports **5**, 302 (1986).
104. – – Production of Indole Alkaloids by Gel-trapped Cells of *Catharanthus roseus* in a Continuous Flow Reactor. Biotechnol. Lett. **8**, 863 (1986).
105. BRODELIUS, P., B. DEUS, K. MOSNACH, and M.H. ZENK: Immobilized Plant Cells

for the Production and Transformation of Natural Products. FEBS Lett. 103, 93 (1979).
106. KUTNEY, J.P., C.A. BOULET, L.S.L. CHOI, W. GUSTOWSKI, M. MCHUGH, J. NAKANO, T. NIKAIDO, H. TSUKAMOTO, G.M. HEWITT, and R. SUEN: Alkaloid Production in *Catharanthus roseus* (L.) G. Don Cell Cultures. XV. Synthesis of Bisindole Alkaloids by Use of Immobilized Enzyme Systems. Heterocycles 27, 621 (1988).
107. STUART, K.L., J.P. KUTNEY, T. HONDA, and B.R. WORTH: Intermediacy of 3′,4′-Dehydrovinblastine in the Biosynthesis of Vinblastine-type Alkaloids. Heterocycles 9, 1419 (1978).
108. STUART, K.L., J.P. KUTNEY, and B.R. WORTH: Studies on the Synthesis of Bisindole Alkaloids. XIV. Enzyme Catalysed Formation of Leurosine. Heterocycles 9, 1015 (1978).
109. MCLAUCHLAN, W.R., M. HASAN, R.L. BAXTER, and A.I. SCOTT: Conversion of Anhydrovinblastine to Vinblastine by Cell-free Homogenates of *Catharanthus roseus* Cell Suspension Cultures. Tetrahedron 39, 3777 (1983)
110. KUTNEY, J.P., B. AWERYN, L.S.L. CHOI, P. KOLODZIEJCZYK, W.G.W. KURZ, K.B. CHATSON, and F. CONSTABEL: Alkaloid Production in *Catharanthus roseus* Cell Cultures. XI. Biotransformation of 3′,4′-Anhydrovinblastine to Other Bisindole Alkaloids. Helv. Chim. Acta 65, 1271 (1982).
111. KUTNEY, J.P., L.S.L. CHOI, T. HONDA, N.G. LEWIS, T. SATO, K.L. STUART, and B.R. WORTH: Biosynthesis of the Indole Alkaloids. Cell-free Systems from *Catharanthus roseus* Plants. Helv. Chim. Acta 65, 2088 (1982).
112. ENDO, T., A. GOODBODY, J. VUKOVIC, and M. MISAWA: Biotransformation of Anhydrovinblastine to Vinblastine by a Cell-Free Extract of *Catharanthus roseus* Cell Suspension Cultures. Phytochemistry 26, 3233 (1987).
113. MISAWA, M., T. ENDO, A. GOODBODY, J. VUKOVIC, C. CHAPPLE, L. CHOI, and J.P. KUTNEY: Synthesis of Dimeric Indole Alkaloids by Cell Free Extracts from Cell Suspension Cultures of *Catharanthus roseus*. Phytochemistry 27, 1355 (1988).
114. ENDO, T., A. GOODBODY, J. VUKOVIC, and M. MISAWA: Enzymes from *Catharanthus roseus* Cell Suspension Cultures that Couple Vindoline and Catharanthine to Form 3′,4′-Anhydrovinblastine. Phytochemistry 27, 2147 (1988).
115. GOODBODY, A.E., T. ENDO, J. VUKOVIC, J.P. KUTNEY, L.S.L. CHOI, and M. MISAWA: Enzymic Coupling of Catharanthine and Vindoline to Form 3′,4′-Anhydrovinblastine by Horseradish Peroxidase. Planta Med. 54, 136 (1988).
116. KUTNEY, J.P., B. BOTTA, C.A. BOULET, C.A. BUSCHI, L.S.L. CHOI, J. GOLINSKI, M. GUMULKA, G.M. HEWITT, G. LEE, M. MCHUGH, J. NAKANO, T. NIKAIDO, J. ONODERA, I. PEREZ, P. SALISBURY, M. SINGH, R. SUEN, and H. TSUKAMOTO: Alkaloid Production in *Catharanthus roseus* (L.) G. Don Cell Cultures. XVI. Biotransformation of 3′,4′-Anhydrovinblastine with *Catharanthus roseus* Cell Cultures and Enzyme Systems. Heterocycles 27, 629 (1988).
117. KUTNEY, J.P., L.S.L. CHOI, J. NAKANO, and H. TSUKAMOTO: Flavine Coenzyme Mediated Photooxidation of 3′,4′-Anhydrovinblastine. Further Information on the Later Stages of Bisindole Alkaloid Biosynthesis. Heterocycles 27, 1927 (1988).
118. – – – – Biomimetic Chemical Transformation of 3′,4′-Anhydrovinblastine to Vinblastine and Related Bisindole Alkaloids. Heterocycles 27, 1937 (1988). See also KUTNEY, J.P., C.A. BOULET, L.S.L. CHOI, W. GUSTOWSKI, M. MCHUGH, J. NAKANO, T. NAKAIDI, H. TSUKAMOTO, G.M. HEWITT, and R. SUEN: Alkaloid Production in *Catharanthus roseus* (L.) G. Don Cell Cultures. XIV. The Role of Unstable Dihydropyridinium Intermediates in the Biosynthesis of Bisindole Alkaloids. Heterocycles 27, 613 (1988).
119. STUART, K.L., J.P. KUTNEY, T. HONDA, N.G. LEWIS, and B.R. WORTH: The Biosynthesis of Vindoline Using Cell Free Extracts from Mature *Catharanthus roseus* Plants. Heterocycles 9, 647 (1978).

120. STÖCKIGT, J., H. GUNDLACH, and B. DEUS-NEUMANN: Disproof of the Overall Enzymatic Biosynthesis of Vindoline from Tryptamine and Secologanin by Cell-Free Extracts from the Leaves of *Catharanthus roseus* (L.) G. Don. Helv. Chim. Acta **68**, 315 (1985).
121. NOE, W., C. MOLLENSCHOTT, and J. BERLIN: Tryptophan Decarboxylase from *Catharanthus roseus* Cell Suspension Cultures: Purification, Molecular and Kinetic Data of the Homogeneous Protein. Plant Mol. Biol. **3**, 281 (1984).
122. PFITZNER, U., and M.H. ZENK: Bound, Stabilized, Highly Pure Strictosidine Synthase and Its Use in the Synthesis of 3α(S)-Strictosidine. Ger. Offen. DE 3, 234, 332.
123. PFITZNER, U., and M.H. ZENK: Immobilization of Strictosidine. Synthase from *Catharanthus* Cell Cultures and Preparative Synthesis of Strictosidine. Planta Med. **46**, 10 (1982).

(*Received December 1, 1988*)

Sucrose and Its Derivatives

By CATHERINE E. JAMES and LESLIE HOUGH, Department of Chemistry, King's College London, University of London, London, U.K., and RIAZ KHAN, Tate and Lyle Research and Technology, Philip Lyle Memorial Research Laboratory, Whiteknights, Reading, Berkshire, U.K.

Contents

1. Introduction 118
   1.1. History 118
   1.2. Structure 119
   1.3. Nomenclature 121
   1.4. Synthesis 122
   1.5. Biosynthesis 125
   1.6. Conformation 127
2. Protective and Functional Groups 129
   2.1. Ethers 129
      2.1.1. Tritylation 129
      2.1.2. Methylation 130
      2.1.3. Silylation 131
   2.2. Cyclic Acetals 132
   2.3. Esters 134
      2.3.1. Carboxylates 135
         2.3.1.1. Acetates 135
         2.3.1.2. Benzoates 138
         2.3.1.3. Benzoylpropionates 139
         2.3.1.4. Pivalates 139
      2.3.2. Sulphonates 140
      2.3.3. Chlorosulphates 143
      2.3.4. Other Esters 148
3. Derivatives 149
   3.1. Anhydrides (Oxetanes) and Epoxides (Oxiranes) 149
   3.2. Halides 154
   3.3. Unsaturated, Deoxy and Branched-Chain Compounds 157
   3.4. Nitrogen-containing Compounds: Azides, Amines and Morpholines 159
   3.5. Sulphur Derivatives 163
4. Enhancement of Sweetness: Structure-Activity Relationships 165

5. Natural Products Containing Sucrose 168
   5.1. β-D-Fructofuranosyl Derivatives 168
   5.2. α-D-Glucopyranosyl Derivatives 169
   5.3. α- and β-D-Galactopyranosyl Derivatives 170
   5.4. Galloyl Derivatives 171
   5.5 Agrocinopine A 172
   5.6. Sucrose Esters from Potato 173
   5.7. Sucrose Esters from Tobacco 173

References 175

1. Introduction

1.1. History (1, 2)

Sucrose, the most abundant of all sugars, has been produced as a food and sweetener since 2000 BC from the juice in the stem of the sugar cane plant, a perennial tropical grass that originated in the South Pacific. As early as 325 BC sugar cane was cultivated in India from where it spread to the Mediterranian countries, and consequently sugar had gradually replaced honey as the major sweetener in Europe by the 15th century. In 1493 Christopher Columbus took sugar cane from the Canary Islands on his second transatlantic voyage to the Caribbean, the old Spanish Main, thus initiating a vast agricultural industry in the tropics and with it the hideous slave trade for plantation labour. In the mid-16th century sugar was a luxury and cost as much as caviar does today.

In 1747 Marggraf discovered that beetroot was a potential source of sucrose. Based on this discovery, Achard in 1801 devised and constructed the first successful factory process for beet sugar refinement at Cunera in Silesia, but it was subsequently destroyed by Napoleon's army. During the Napoleonic wars the blockade of its ports deprived Europe of cane sugar from the East und West Indies, this resulted in an extensive search for alternative sweeteners which was encouraged by prizes and awards from Napoleon Bonaparte. Thus, in 1892, Proust isolated glucose from grapes and in 1811 Kirchoff converted starch into glucose by acid hydrolysis. The latter process led to the creation of a huge industry based on corn, wheat and potatoes although the hydrolysis is now catalysed by enzymes. To Napoleon's obvious delight sucrose was manufactured in 1803 by Benjamin Delessert from sugar beet at Plassy near Paris. This process, created in Europe during the early 1800's, rapidly spread through Russia, Denmark, Holland, the U.S.A. and Canada and is now the predominant source of sugar in

References, pp. 175–184

Europe. White beetroot (*beta vulgaris*) yields 10–20% by weight of sucrose and is an annual plant. Sugar from beet now represents about 40% of the total world production of sucrose which in 1988 exceeded 100 million tons. The remaining 60% of world production is manufactured from the stem of the sugar cane, the richest source being *Saccharum officinarum* whose expressed jurce usually contains 20% by weight of sucrose. Sugar production has increased enormously during the 20th century, exhibiting more than a tenfold increase since the turn of the century, and as a leading world commodity it is certainly the most readily available, lowest cost crystalline organic compound.

1.2. Structure (*3*)

Sucrose ($C_{12}H_{22}O_{11}$) exists in two crystalline forms, the normal stable sucrose A with m.p. 184–185° from water and a metastable sucrose B with m.p. 169–170°, obtained by recrystallisation from methanol. It is a non-reducing disaccharide with $[\alpha]_D + 66.53°$ (H_2O) which on hydrolysis with very dilute acid, cationic exchange resin or the enzyme invertase yields equimolar proportions of D-glucose and D-fructose. The hydrolysate is laevo-rotatory ($[\alpha]_D - 28.2°$); consequently the process became known as 'inversion', and the product as 'invert sugar'. Acetylation of sucrose yields an octa-acetate, reflective of the eight hydroxyls present, with $[\alpha]_D + 59.6°$ ($CHCl_3$) and existing in two crystalline forms with m.p. 69–70° and 75° respectively.

The ring structure present in sucrose mystified a number of great organic chemists since at least four different structures were proposed containing from three to seven membered rings (Fig. 1.). In 1883 Tollens proposed a glucoseptanosyl fructofuranoside (**1**), in 1892 Fischer concluded that sucrose was a glucofuranosyl fructofuranoside (**2**) whilst in 1930 HUDSON (**4**) favoured a glucopyranosyl fructo-oxetanoside (**3**) since he did not accept the correct structure proposed-earlier, in 1927, by HAWORTH and HIRST (*5, 6*), namely a glucopyranosyl fructofuranoside (**4**). They had deduced as a result of their classical methylation study, which resulted in isolation of 1,3,4,6-tetra-*O*-methyl-D-fructose and 2,3,4,6-tetra-*O*-methyl-D-glucose from the hydrolysis of octa-*O*-methyl sucrose. Structure (**4**) was confirmed by Malapradian periodate oxidation (*7*) of sucrose which consumed 3 mol. of periodate and formed one mol.-equivalent of formic acid, but significantly no formaldehyde, and also produced 'tetra-aldehyde' which was characterised.

Finally, the α- and β-configurations at the two glycosidic centres in sucrose, the anomeric carbon-1 and carbon-2' of the glucose and fructose units respectively, were established as a result of the hydrolysis

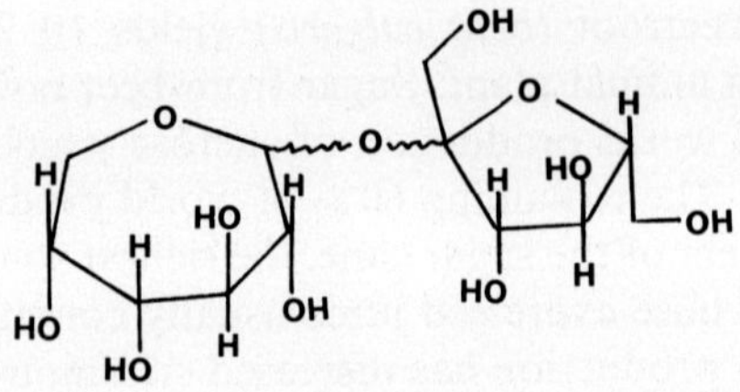

Fig. 1. Chemical structures suggested for sucrose

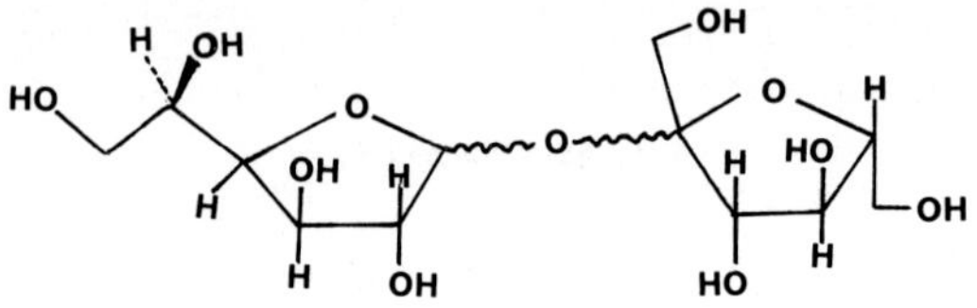

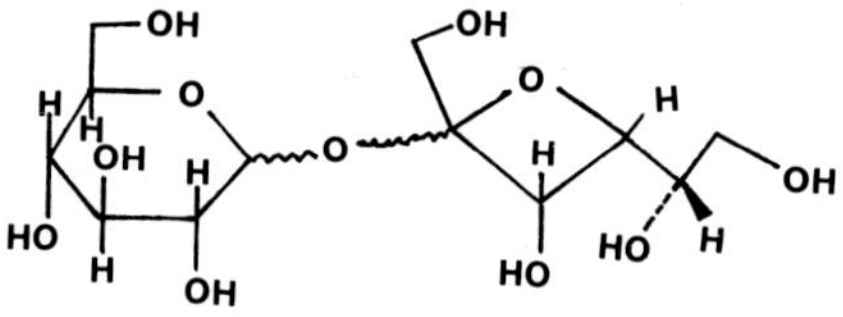

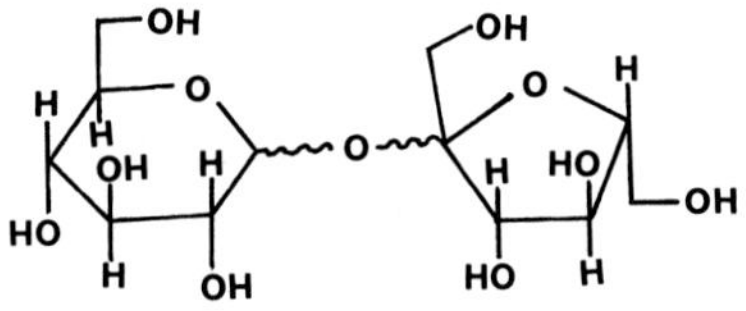

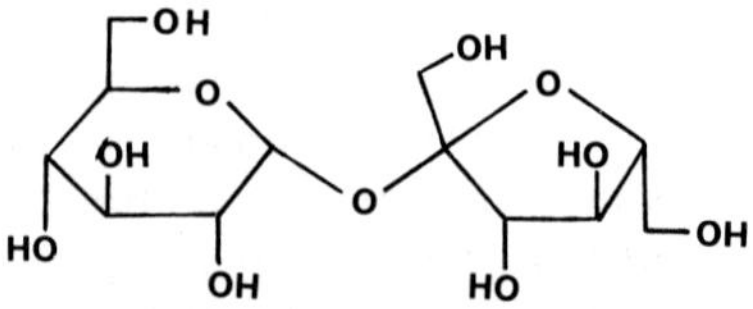

catalysed by the specific enzymes α-D-glucosidase and β-D-fructofuranosidase (invertase) (*8*) and by X-ray investigations (*9*) of sodium chloride and sodium bromide addition compounds, thereby establishing sucrose as α-D-glucopyranosyl β-D-fructofuranoside (**5**). It contains 9 chiral centres and is unique in that of the 512 possibilities only the one stereoisomer (**5**) exists in Nature.

The facile hydrolysis of sucrose (**5**) (Fig. 2.) is explained by the collapse of the protonated form, the oxonium ion (**6**), to give D-glucose and the D-fructofuranosyl carbo-cation (**7**), a tertiary carbonium ion, which is then fransformed into D-fructose (**8**). In accord with this mechanism sucrose is cleaved by 0.1% methanolic hydrogen chloride at 20° to give D-glucose and methyl D-fructofuranosides (**9**) in 34 minutes (*10*).

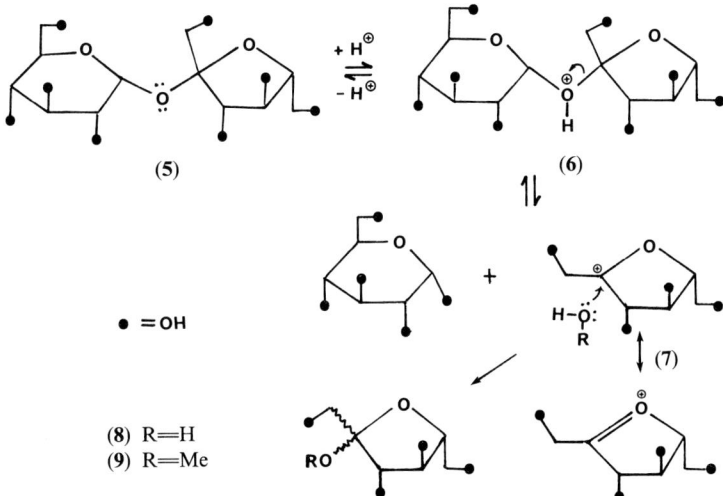

Fig. 2. Acid hydrolysis of sucrose

1.3. Nomenclature

The term glycosyl denotes a radical derived from a sugar by detaching the hemiacetal or anomeric hydroxyl group. Consequently non-reducing disaccharides are glycosyl glycosides and sucrose can be termed either α-D-glucopyranosyl β-D-fructofuranoside or β-D-fructofuranosyl α-D-glucopyranoside.

When the trivial name is used unprimed and primed numbers are used to indicate carbon, and attached oxygen, atoms in the glucosyl and fructoside units respectively (**5**).

If, for example, 4-hydroxyl group of sucrose is inverted the product is known either by the trivial name *galacto*sucrose or the more correct α-D-galactopyranosyl β-D-fructofuranoside.

The chemistry of sucrose has been reviewed by JENNER (*11*) and by KHAN (*12*).

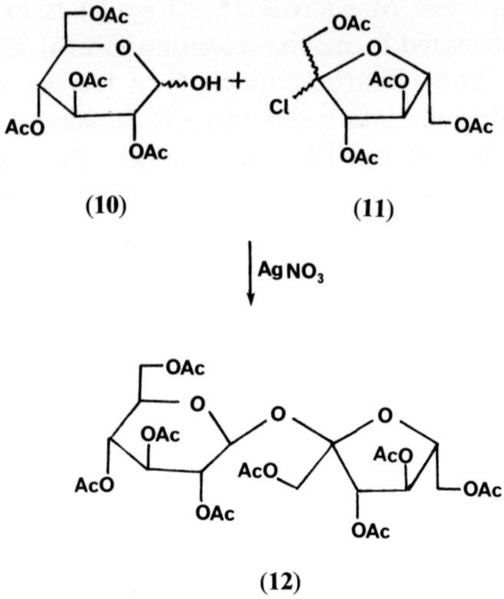

Fig. 3. Synthesis of *iso*-sucrose (β-D-glucopyranosyl α-D-fructofuranoside)

1.4. Synthesis

Many early attempts (*13*) to synthesise sucrose by traditional methods each gave "iso-sucrose" octa-acetate (**12**) (Fig. 3.) with the unwanted, unnatural β-D-glucosyl and α-D-fructoside linkages, since there was no steric control of the anomeric configurations at the two glycosyl units. All of the reactions involved 2,3,4,6-tetra-*O*-acetyl-α-D-glucopyranosyl derivatives, for example its 1-hydroxy derivative (**10**) with 1,3,4,6-tetra-*O*-acetyl-D-fructofuranosyl chloride (**11**), and participation by the neighbouring 2-acetoxy group probably guides the synthesis to the β-glucoside (**12**). In 1953, LEMIEUX and HUBER (*14*) overcame this problem by using 3,4,6-tri-*O*-acetyl-1,2-anhydro-α-D-glucopyranose (**13**), known as Brigl's anhydride, which reacted with 1,3,4,6-tetra-*O*-acetyl-D-fructofuranose (**14**) in benzene at 100 °C for 104 hours, to give sucrose octa-acetate (**15**) in 5.5% yield (Fig. 4.). The α-glucoside (**15**) resulted from participation by the 6-acetoxy group in the reaction in the absence of a neighbouring 2-acetoxy group. A fusion reaction

References, pp. 175–184

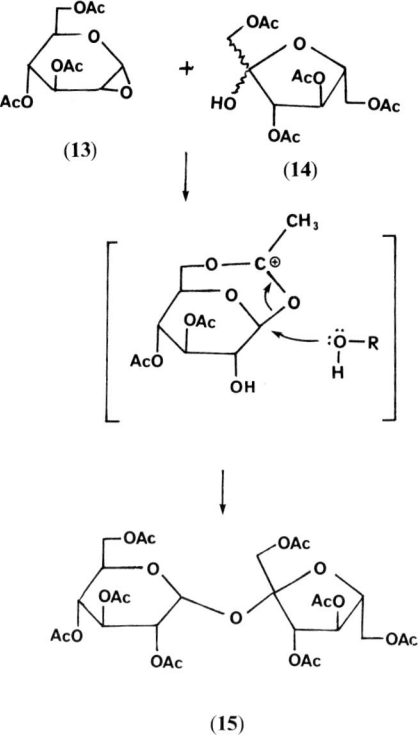

Fig. 4. Synthesis of sucrose from Brigl's anhydride

of Brigl's anhydride (13) with 1,3,4,6-tetra-O-benzoyl-D-fructofuranose at 110–118 °C for 3 hours also gave sucrose but in similar yield (15).

Another approach to synthesis to avoid participation by neighbouring groups involved condensation of 2,3,4,6-tetra-O-benzyl-α-D-glucopyranosyl chloride with 1,3,4,6-tetra-O-benzyl-D-fructofuranose which gave a crude syrupy octa-O-benzylsucrose in 36% yield (16). A similar synthesis (17) of L-sucrose, the mirror image of natural D-sucrose, employed the benzyl ethers of L-glucose (prepared from L-arabinose) and L-fructose (made from L-mannitol). The debenzylated product L-sucrose was similar in sweetness to the natural isomer. Significantly, L-glucose and L-fructose are also similar in sweetness to their natural D-counterparts but in the case of amino acids, the D-isomers are usually sweet whereas the L-isomers are bitter (see Section 4.). A novel synthesis of sucrose, also generally applicable to α-linked oligosaccharides (18), utilised a dienoid receptor (16) which reacted with 1,3,4,6-tetra-O-acetyl-α,β-D-fructofuranose (**14**) to give the unsaturated α-glucoside (**17**) when treated with the iodonium ion (Fig. 5.). Sequential hydroxylation, selective benzoylation, oxidation, reduction and acetylation gave a mixed product from which 4,6-*O*-benzylidene sucrose hexa-acetate (**18**) was isolated (*18*).

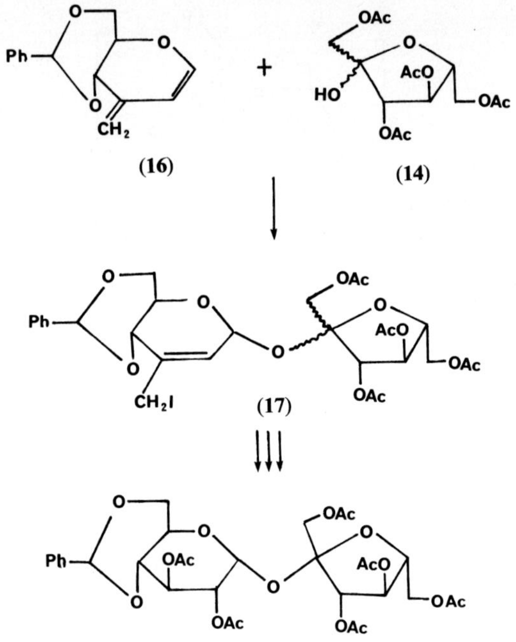

Fig. 5. Synthesis of sucrose starting from a dienoid receptor

Interest in the mechanism of sweetness initiation and reception and in the design of non-nutritive sweeteners led to syntheses of *C*-sucrose (**24**) and *S*-sucrose (**27**). The key intermediates in the synthesis (*19*) of *C*-sucrose were the vinyl iodide of the *C*-α-glucoside of 2,3,4,6-tetra-*O*-benzyl-D-glucopyranoside (**19**) and 2-D-benzyl-3,4-*O*-isopropylidene-D-erythrose (**20**) which were coupled by Ni(*II*)/Cr(*II*)-mediation to yield the desired *threo*-allylic alcohol (**21**), but as the minor product, together with the major *erythro*-isomer (Fig. 6.). Epoxidation of the former (**21**) with *m*-chlorobenzoic acid gave a 1:3 mixture of syn- and anti-epoxy alcohols, then opening of the *threo*-anti epoxide (**22**) with camphor sulphonic acid and concomittant cyclisation gave the required triol

(23) which upon hydrogenolysis of the benzyl ether groups afforded the acid-stable C-sucrose (24). The flexibility of this synthesis made the C-2' and C-3' stereoisomers of C-sucrose available for sensory studies.

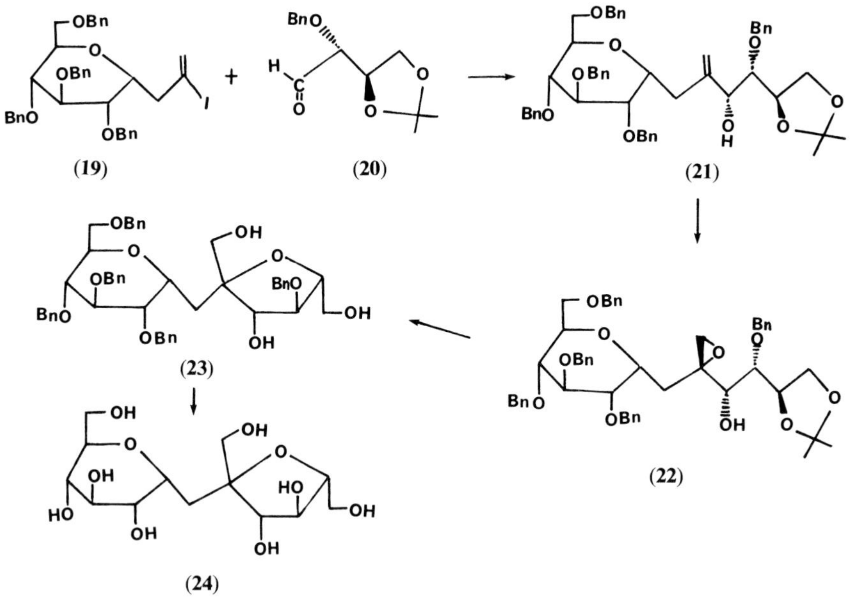

Fig. 6. Synthesis of C-sucrose

The synthesis (20) of the 1-thio-α-D-glucosyl analogue of sucrose (27) and its α,α-anomer (28), was achieved by a Lewis acid-catalysed condensation between 2,3,4,6-tetra-O-acetyl-1-thio-α-D-glucose (25) and 1,3,4,6-tetra-O-benzyl-D-fructofuranose (26). Deacetylation and debenzylation followed by chromatography afforded 26 in 15% yield and α-D-fructofuranosyl 1-thio-α-D-glucopyranoside (27) in 22% yield (Fig. 7.). Such compounds are of biological interest because S-sucrose (27) induces biosynthesis of the levansucrase of *Bacillus subtilis*, an enzyme which catalyses the transfructosylation from sucrose to a suitable acceptor via a fructosyl-enzyme intermediate (20). Furthermore it acts as a competitive inhibitor for both levansucrase and invertase.

1.5. Biosynthesis

In higher plants sucrose appears to arise by two pathways. Common to each of these is the enzymic transfer of a glucosyl unit from uridine

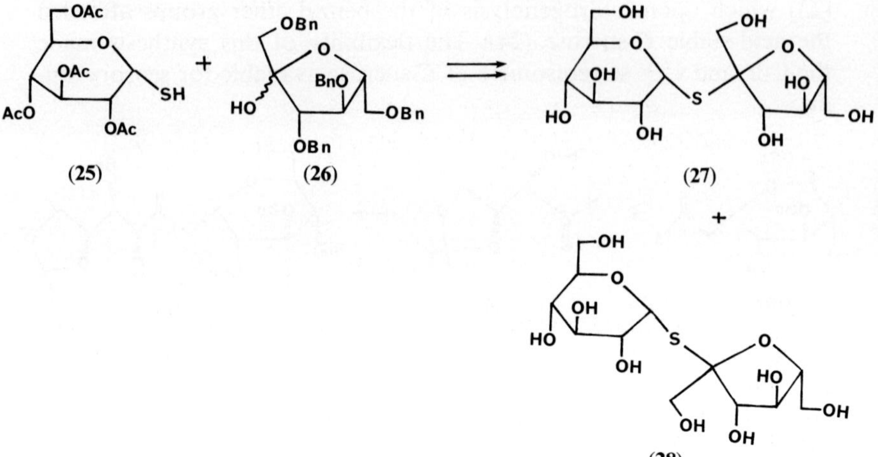

Fig. 7. Synthesis of S-sucrose (27)

diphosphate glucose (UDP) to either fructose or fructose 6-phosphate. The second pathway leading to sucrose 6'-phosphate would appear to be the more important and the product is then deesterified by a phosphatase to give free sucrose (21, 22).

A phosphorylase enzyme, such as that from *Pseudomonas saccharophila*, catalyses the transfer of glucose from α-D-glucosyl 1-phosphate ("Cori ester") to D-fructose to give sucrose (3, 23), with the release of inorganic phosphate. The enzyme is not specific in terms of the ketose acceptor and substitution of D-fructose by D-*threo*pentulose (D-xylulose) and by L-sorbose led to the isolation of the unusual non-reducing disaccharides α-D-glucopyranosyl β-D-*threo*pentuloside and α-D-glucopyranosyl α-L-sorbofuranoside respectively (24).

In seeking substituted sucrose derivatives that are not substrates for invertase but which are transported in the plant by sucrose specific carriers – proteins in the membranes of leaf vascular tissue – CARD and HITZ (25) synthesised 1'-deoxy-1'-fluorosucrose by the action of a sucrose synthetase (D-fructose D-glucosyl transferase) from barley seeds upon UDP-glucose and 1-deoxy-1-fluoro-D-fructose. Sucrose synthetase is remarkably unspecific in its action since various deoxy-D-fructoses can act as receptors for a selection of UDP-deoxyglucose donor substrates, thus permitting the synthesis of a variety of deoxy-sucroses, such as 1-deoxy-, 4-deoxy- and 6-deoxy-β-D-fructofuranosyl α-D-glucopyranosides, 6,6'-dideoxysucrose, 2-deoxy-, 3-deoxy-, 4-deoxy- and 6-deoxy-α-D-glucopyranosyl β-D-fructofuranosides (26).

References, pp. 175–184

However the affinity of the enzyme and the rate of hexose incorporation decreases when deoxy sugars are utilised.

On the other hand invertase (β-D-fructofuranosidase) is highly specific and none of the deoxy-β-D-fructosyl derivatives are hydrolysed, but α-glucosidase (α-D-glucopyranosidase) is not as specific, since it hydrolysed all of the above deoxy sucroses (*26*).

1.6. Conformation

The conformations of the rings of the sucrose molecule were determined by use of neutron diffraction techniques (*27*). It revealed in the crystal form two strong intramolecular hydrogen bonds between O-2 and HO-1' and between O-5 and HO-6' (Fig. 8.). These bonds serve to "hold" the molecule in a well-ordered, rigid conformation (**29**) in which the two rings are approximately at right angles and in which the pyranosyl chair adopts a $^4C_1$ conformation whilst that adopted by the furanoside ring is a $_3T^4$ twist. Apart from the 4-hydroxyl group the other hydroxyls are all intermolecularly hydrogen bonded. The conformation in aqueous solution however has been the subject of much debate. BOCK and LEMIEUX (*28*) predicted by use of hard sphere calculations that dilute *aqueous* solutions of sucrose had a similar conformation to the crystal structure but with only one intramolecular H-bond, from O-2 to HO-1' (**30**) and they obtained supporting evidence from ^{1}H- and ^{13}C-nmr spectral data of D_2O and $(CD_3)_2SO$ solutions. Using Xray and Raman techniques, MATHLOUTHI *et al.* (*29*) found no evidence of intramolecular H-bonds in dilute solution (< 0.7 M) but as the concentration was increased to a saturated solution intramolecular H-bonds were formed, resembling those in the crystal structure. The results were interpreted to show that the conformation of aqueous sucrose was dependent upon concentration but to the contrary MCCAIN and MARKLEY (*30*), using more reliable ^{13}C-nmr spin lattice relaxation measurements, showed that the conformation is independent of both temperature and concentration (from 0.1 M to 1.0 M) and confirmed the results of BOCK and LEMIEUX (*28*).

Application of secondary isotope multiplet partially labelled entities (SIMPLE) nmr spectroscopy to solutions of sucrose in deuterated dimethylsulphoxide solution (*31*) revealed the presence of two different conformations with intramolecular hydrogen bonds from either the 1'-hydroxyl or the 3'-hydroxyl of the fructofuranoside as donors to the acceptor 2-hydroxyl of the glucopyranosyl unit (**30** and **31** respectively) (Fig. 8.). These conformations are in competitive equilibrium, with the 1'-OH ″″″″ 2-O predominating over the 3'-OH ″″″″ 2-O, in the

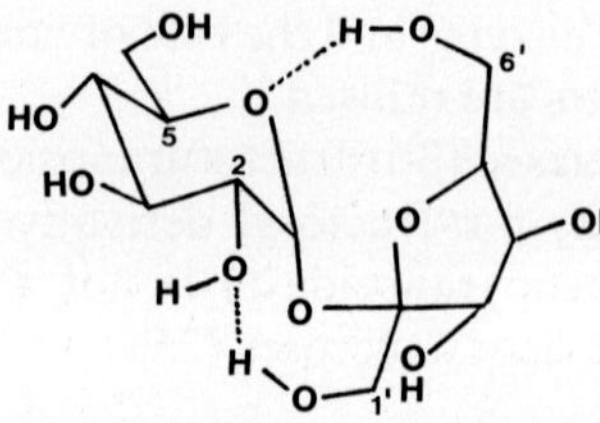

Fig. 8. Conformation of sucrose in the crystal (**29**) and in solution (**30**, **31**)

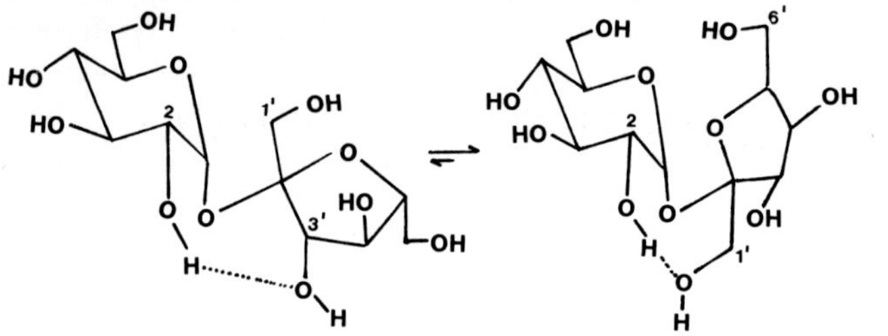

ratio of 2:1 respectively. Furthermore, the inter-unit hydrogen bond is strengthened by co-operative bonding from neighbouring hydroxyl groups in both glucosyl and fructoside units. The magnitudes of the torsion angles between the inter-unit bonds reflect the relative disposition of the two rings. The torsion angles (O-5 – C-1 – O-1 – C-2′) and (C-1 – O-1 – C-2′ – O-2′) are 107.6° and −44.4° respectively in the crystal.

The infrared spectrum of sucrose shows a strong vibrational frequency at 3560 cm^{-1} characteristic of a free hydroxyl group and assigned to the equatorial HO-4 (*32*). This complements neutron diffraction data (*27*) of sucrose which suggests that it is not engaged in hydrogen bonding. No such frequency is seen in the spectrum of *galacto*sucrose.

The intrinsic viscosity of sugars reflects the extent of hydration which is greatly influenced by the conformation of the molecules and in particular by the configuration of the hydroxyl (OH) groups. Since equatorial-OH groups are more readily hydrated than axial-OH groups, the higher the number of equatorial hydroxyl groups in a sugar molecule, the higher the intrinsic viscosity. This explains why maltose has

References, pp. 175–184

a higher intrinsic viscosity than sucrose and lactose which exhibit approximately the same intrinsic viscosity (*33, 34*). The dynamic interplay of hydrogen bonds in aqueous solutions of sugars governs their solution properties and probably contributes to their taste qualities. Sucrose is widely used as a humectant in the food industry and for the control of rheology. An ^{17}O-nmr spin lattice and transverse relaxation time study (*35*) of aqueous solutions (0–70% w/w) led to the conclusion that each sucrose molecule exists as a hydrated monomer and slightly modifies the correlation time of *ca.* 16 water molecules loosely bound to it at any one time.

2. Protective and Functional Groups

2.1. Ethers

2.1.1. Tritylation

The triphenylmethyl or trityl group has found extensive use as a protecting group for primary hydroxyl groups in carbohydrate synthesis (*36*). Trityl ethers of sucrose are conveniently prepared by treatment with approximately stoichiometric amounts of chlorotriphenylmethane in dry pyridine at room temperature or above. These compounds are commonly crystalline solids which are resistant to base and other nucleophiles but which are readily hydrolysed under weakly acidic conditions such as boiling aqueous acetic acid or HBr in glacial acetic acid and by hydrogenolysis.

The first synthesis of 6,1',6'-tri-*O*-tritylsucrose in crystalline form was reported by JOSEPHSON in 1929 (*37*) but his results could not be reproduced by later workers (*38*). In 1957, the preparation of 6,1',6'-tri-*O*-tritylsucrose penta-acetate by treatment of sucrose with 3.5 molar equivalents of trityl chloride in pyridine followed by addition of acetic anhydride to the reaction mixture was reported by McKEOWN *et al.* (*38, 39*).

In 1970 OTAKE (*40*) established that the order of reactivity of the primary hydroxyl groups in sucrose towards tritylation is C-6 ≃ C-6' > C-1'. This conclusion arose from his study of the reaction of sucrose with 1.2 molar equivalents of trityl chloride in pyridine at ambient temperature for 96 hr which afforded 6- and 6'-*O*-tritylsucroses in 18 and 19% yields but none of the 1'-*O*-tritylsucrose. In addition, the dimolar tritylation of sucrose (*41*) yielded 6,6'-, 6,1'- and 1',6'-di-*O*-

tritylsucrose in yields of 27, 4 and 5% respectively. Further evidence was obtained from the reaction of sucrose with four molar equivalents of trityl chloride which gave 6,6'-di-tritylsucrose and 6,1',6'-tri-O-tritylsucrose in 30 and 68% yields respectively (*42*).

In 1972, 1',6'-di-O-tritylsucrose hexa-acetate was isolated in 55% yield by BUCHANAN and CUMMERSON (*43*) from 2,3,6,3',4'-penta-O-acetylsucrose by reaction with 1.9 molar equivalents of trityl chloride in pyridine at 100 °C for 20 h followed by conventional acetylation. These workers also achieved selective tritylation by treatment of the above penta-O-acetylsucrose with 1.6 molar equivalents of trityl chloride in pyridine at 50 °C for 32 h. After de-esterification, 6'-O-tritylsucrose was obtained in 52% yield (*44*).

An interesting synthesis of 1'-O-tritylsucrose in 21% yield was reported by OTAKE in 1974 (*45*). This was achieved *via* the tritylation of commercially available 6,6'-di-O-palmitoylsucrose followed by de-esterification of the product. KHAN and MUFTI (*46*) synthesised the 6,1'-di-O-tritylsucrose hexa-acetate in >70% yield in a similar way from 3,3',4',6'-tetra-O-acetylsucrose using 2 molar equivalents of trityl chloride in pyridine at 90 °C for 24 h, chromatographic separation and conventional acetylation.

2.1.2. Methylation

Exhaustive methylation has found extensive application in the structural determination of carbohydrates including sucrose (*3, 6*). Initially etherification of sucrose occurs preferentially at the most acidic hydroxyl groups (0–1', 0–2 and 0–3'), followed by the least hindered hydroxyl groups (0–6 and 0–6'). The preparation of sucrose methyl ethers has been achieved by use of a variety of reagents, including dimethyl sulphate-sodium hydroxide (*47, 48*), methyl iodide-silver oxide-acetone, sodium hydride-methyl iodide-N,N-dimethylformamide and diazomethane-boron trifluoride etherate (*49*). Partially methylated sucrose derivatives have been synthesised for taste evaluation in order to identify those hydroxyls that are responsible for the sweet taste of sucrose (*11, 49*) (see section 3.0). The diazomethane method finds particular application in the methylation of partially esterified sucrose derivatives on account of the mild conditions involved which allow the reaction to occur without causing acyl group migration or hydrolysis. Thus treatment of the hexa-O-acetyl sucroses with diazomethane-boron trifluoride etherate in dichloromethane at −5 °C for 30 min afforded the corresponding di-O-methyl ethers, 6,6'-di-, 1',6'-di-, and 4,6-di-O-methyl sucrose respectively and no ester migration was observed (*43,*

49). Controlled etherification of sucrose using sodium hydroxide and dimethyl sulphate has been observed to achieve partial alkylation, yielding the 3'-O-methyl and 4-O-methyl derivatives. The latter product occurred as the major mono-O-methyl derivative due to its resistance to further methylation when compared with other mono-O-methyl sucroses (*50, 51*).

2.1.3. Silylation

Per-O-trimethylsilyl (tms) ethers of sucrose are prepared by reaction with chlorotrimethylsilane in either hexane (*52*) or pyridine (*53*), but this reaction appears to be non-selective. Crystalline octa-O-(trimethylsilyl) sucrose has been isolated from the reaction of sucrose with hexamethyldisilazane and chlorotrimethylsilane in pyridine (*50*). Trimethylsilylation renders sucrose sufficiently volatile to enable analysis by g.l.c. which has been often used for its assay in the presence of other carbohydrates.

Hydroxyl groups of sucrose may be protected by means of the *t*-butyldimethylsilyl group. Whilst being stable to normal alkylation and acylation reactions, this type of silyl ether is readily hydrolysed by tetrabutylbutylammonium fluoride giving the parent hydroxy compound (*55*).

Treatment of sucrose with *t*-butyldimethylsilyl chloride in pyridine gives rise to a number of useful partially protected derivatives and the reagent shows selectivity toward primary hydroxyl groups (*56*). Thus reaction of sucrose with 3.5 molar equivalents of *t*-butyldimethylsilyl chloride in pyridine afforded the 6,1',6'-tri-O-*t*-butyldimethylsilyl derivative in 64% yield. In an analogous reaction using only 0.65 molar equivalents of the chlorosilane, the 6'-mono and 6,6'-di-ethers were obtained in 10 and 36% yields respectively (*56*). The absence of monosubstitution at C-6 and C-1' suggests that HO-6' in sucrose is the most reactive site towards this silylation reaction.

Similar selectivity was observed in the formation of the *t*-butyldiphenylsilyl ethers of sucrose as for the less hindered *t*-butyldimethylsilyl ethers. However, the *t*-butyldiphenylsilyl group shows greater stability towards acid and hydrogenolysis than the related silyl and trityl ethers.

6'-O-*t*-Butyldiphenylsilyl sucrose was obtained in 49% yield from reaction of the appropriate chlorosilane (1.1 equivalent) with sucrose in pyridine in the presence of a catalytic amount of 4-dimethylaminopyridine (*57*). When 3 molar equivalents of the reagent were used, the reaction yielded crystalline 6,6'-di-O-*t*-butyldiphenylsilyl sucrose (78%) and 6,1',6'-tri-O-*t*-butyldiphenylsilyl sucrose (19%). The tri-ether was obtained as the major product when 4.6 molar equivalents of the silylating agent were employed (*57*).

2.2. Cyclic Acetals

Despite their importance as synthetic intermediates, cyclic acetals of sucrose long defied preparation due to the acid lability of the disaccharide in conventional acetalation procedures which gave rise to the formation of acetals of D-glucose and D-fructose. In 1974, KHAN (*58*) circumvented the problem by utilising α,α-dibromotoluene in pyridine at elevated temperature (85 °C) and after acetylation obtained 4,6-*O*-benzylidenesucrose hexa-acetate (**32**) in 35% yield (Fig. 9). Application to sucrose of the novel reagent 2,2-dimethoxypropane in *N,N*-dimethylformamide (DMF) with toluene-*p*-sulphonic acid as catalyst, gave 4,6-*O*-isopropylidene (**33**) and 2,1′:4,6-di-*O*-isopropylidene (**34**) derivatives in good yields (*46, 59*). The unique 8-membered 2,1′-cyclic acetal bridges the two rings in sucrose, is more stable to acid hydrolysis than the 4,6-*O*-acetal (*60*) and has been effective in allowing entry to 2- and 1′-derivatives of sucrose (see section 2.3.3.). Prolonged reaction of the 2,2-dimethoxypropane reagent with sucrose for 36 hours gave several new acetals of D-glucose and methyl α-D-fructofuranoside (*61*),

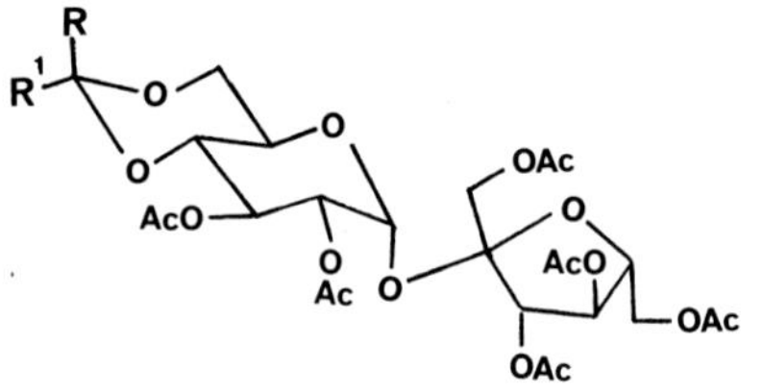

(**32**) R=H, R^1=Ph
(**33**) R=R^1=CH$_3$

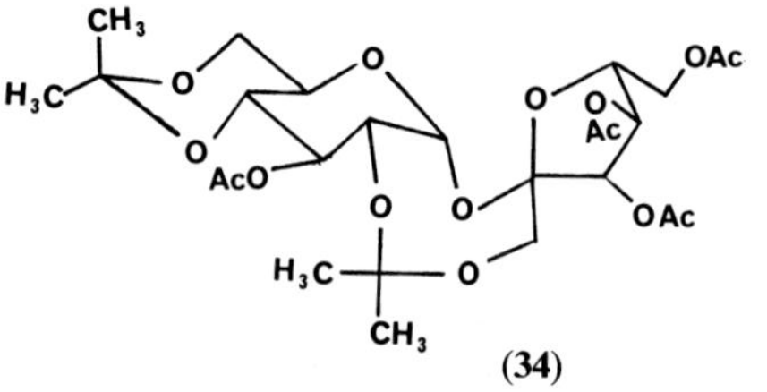

(**34**)

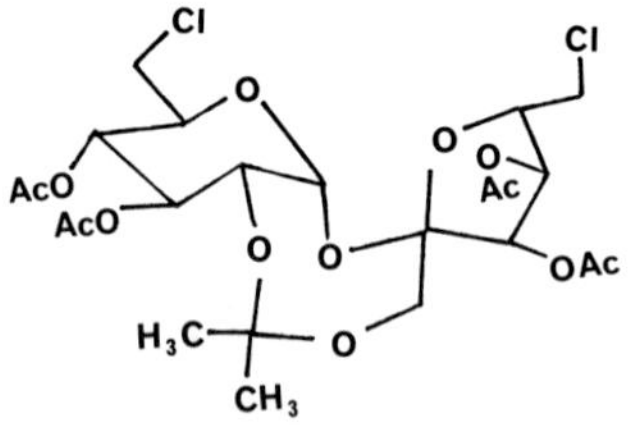

(**38**)

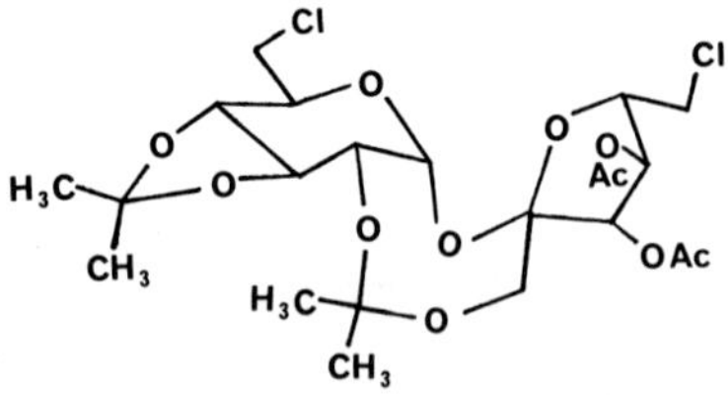

(**39**)

Fig. 9. Cyclic acetals of sucrose and its 6,6′-dichloride

including 1,2:4,6-di-*O*-isopropylidene-α-D-glucopyranose (**36**) and methyl 1,3-*O*-isopropylidene-α-D-fructofuranoside (**37**). The reaction propably proceeds via the 6,1',6'-tri-*O*-(1-methoxy-1-methylethyl)ether (**35**), followed by fission of the C-2'-to-O-2' bond and cyclisation and rearrangement of the acetals (see Fig. 10).

The replacement of 2,2-dimethoxypropane with either cyclohexylidene dimethyl acetal or benzaldehyde dimethyl acetal gave the 4,6-*O*-cyclic acetals (for example **32**) of sucrose exclusively (*11*; *59*).

In the absence of the 6-hydroxy group, as in 6,6'-dichloro-6,6'-dideoxysucrose, the 2,2-dimethoxypropane reaction gave the 2,1'-isopropylidene derivative (**38**), together with the 2,1':3,4-di-*O*-isopropylidene derivative (**39**), isolated in yields of 37% and 40% respectively (*62*) (Fig. 9.). Blockage of all three primary hydroxyl groups by use of trityl groups directed the course of the reaction with the dimethoxypropane reagent to the 2,3-*O*-isopropylidene (24%) and 3,4-*O*-isopropylidene (20%) derivatives (*62*). In general the ease of formation of isopropylidene acetals of sucrose proceeds in the following order of preference: 4,6 > 2,1' > 2,3 > 3,4 (*11*).

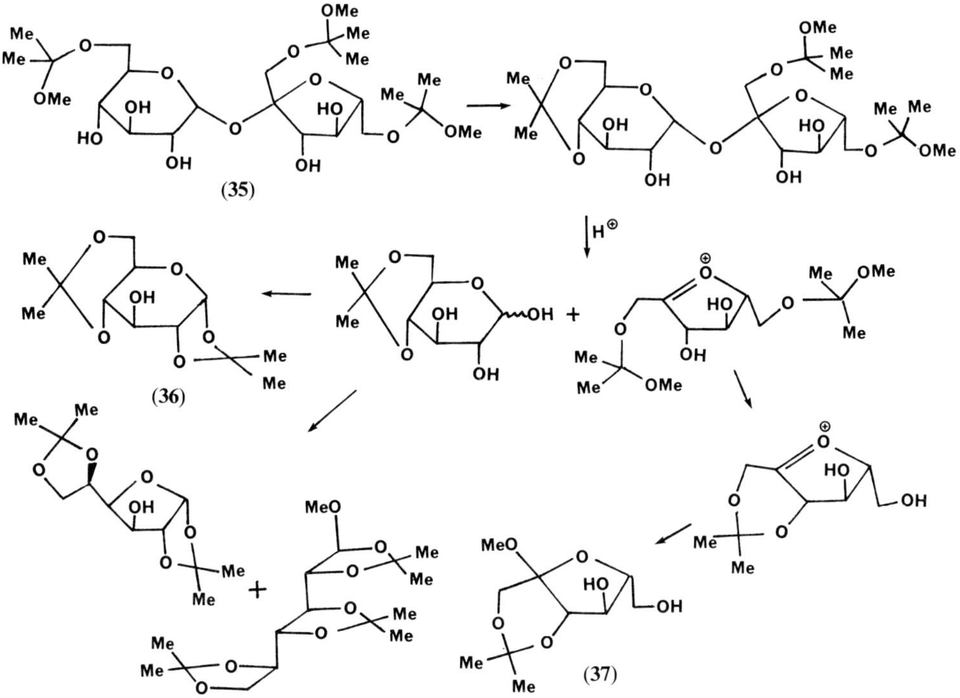

Fig. 10. Acetals of D-glucose and D-fructose derived from sucrose

The phenylsilylene derivatives of sucrose were synthesised by the same approach as above, using dimethoxyphenylsilane in DMF with *p*-toluenesulphonic and afforded the expected eight-membered 2,1'-*O*-silylene acetal (**40**) in addition to a unique 2,1':6,6'-di-*O*-(diphenylsilylene) acetal (**41**), where the 6- and 6'-hydroxyls are spanned by a 12-membered acetal ring (*11*) (Fig. 11.). Apparently the 4,6-*O*-silylene acetal was not formed because this derivative would introduce an axial phenyl group onto the six-membered chair conformation.

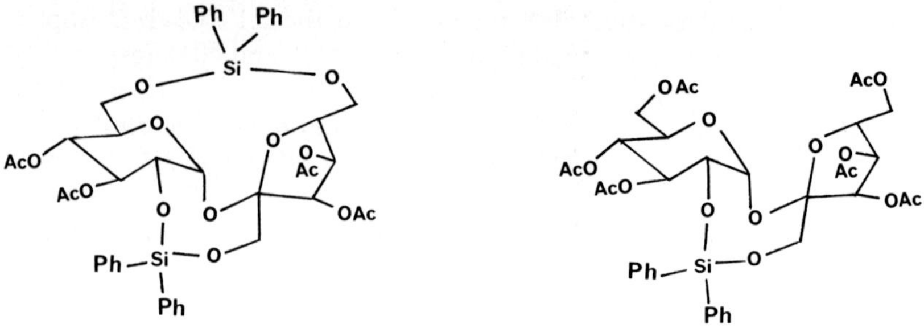

(**40**) (**41**)

Fig. 11. Phenylsilylene derivatives of sucrose

2.3. Esters

Sucrose octa-esters are readily prepared by use of an excess of acylating reagent in pyridine. Sucrose octa-acetate is a well known bittering agent and denaturant and in common with other sucrose esters, it is considerably more stable to acid hydrolysis than the parent compound. The major area of interest is centred on selective esterification reactions.

The eight hydroxyl groups of sucrose exhibit subtle differences in their reactivities towards a variety of acylating agents thus facilitating partial and selective esterifications under carefully controlled reaction conditions. To have available methods for the selective protection of certain hydroxyl groups by this means is of obvious advantage in the chemical manipulation of this molecule. In general, the primary hydroxyls at C-6 and C-6' would be expected to be the most reactive, giving 6- and 6-mono esters and the 6,6'-diester. These reactions will then be followed at the more hindered primary hydroxyl at C-1' leading to the 1',6,6'-triester. Of the five remaining secondary hydroxyls, the

2-hydroxyl and the 3'-hydroxyl would be anticipated to react more readily because of their close proximity to the anomeric centres C-1 and C-2' respectively.

2.3.1. Carboxylates

2.3.1.1. Acetates

Sucrose monoacetate was first prepared in 1865 (63) but more reliable documentation was given by KONENKO and KESTENBAUM (64) who claimed a 95% yield of mono-O-acetylsucrose from reaction of sucrose with 1.6 molar equivalents of acetic anhydride in pyridine. The product, a white crystalline powder, was not fully characterised and almost certainly contained a mixture of regioisomers. In a similar reaction but carried out at $-40\,°C$, KHAN and MUFTI (12, 65) isolated 6-O-acetylsucrose (**42**) in 40% yield after chromatographic separation of the mixture obtained. Characterisation of this compound was supported by ^{13}C-nmr and by chemical transformations.

(42)

The most common routes to partially acylated sucrose derivatives have been *via* tritylether (see Section 2.1.) and cyclic acetal derivatives (see Section 2.2.). In employing this type of procedure, the possibility of acyl group migration must be always considered. Thus detritylation of 6,1',6'-tri-O-tritylsucrose penta-acetate (**43**) with boiling aqueous acetic acid occurs with concomitant migration of the acetyl group from O-4 to O-6 via an 4,6-orthoester intermediate. The product thus obtained is therefore 2,3,6,3',4'-penta-O-acetylsucrose (**44**) (*38*) (Fig. 12). Acetyl migration was minimised in the detritylation procedure by use of HBr in glacial acetic acid and chloroform at $0\,°C$ which gave 2,3,4,3',4'-penta-O-acetylsucrose (**45**) in 79% yield (66). Acyl group migration was not observed under the above conditions during detritylation of the corresponding sucrose pentabenzoate and of 6,6'-ditritylsucrose hexabenzoate (*42*) because the formation of the intermediary cyclic 4,6-orthoester is unfavourable.

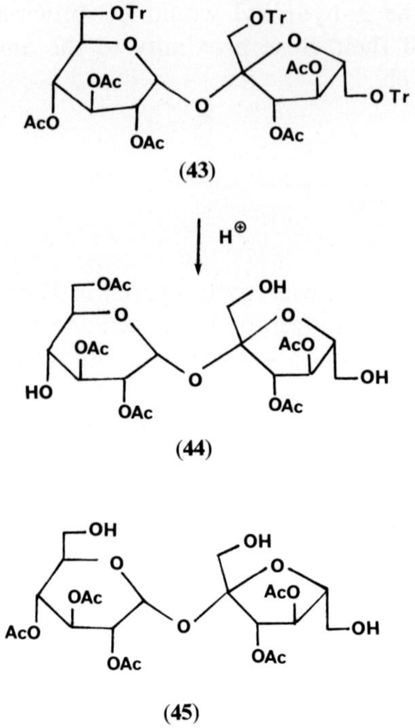

(43)

↓ $H^{\oplus}$

(44)

(45)

Fig. 12. Detritylation with and without ester migration

An alternative route to partially acetylated sucrose derivatives is by the selective de-esterification of sucrose octa-acetate (*67*). This has been achieved by using a column of alumina (Laporte Type H) and chloroform as eluant from which 2,3,4,6,1′,3′,4′-hepta-*O*-acetylsucrose (**46**), 2,3,6,1′,3′,4′,6′-hepta-*O*-acetylsucrose (**47**) and 2,3,4,6,1′,3′,6′-hepta-*O*-acetylsucrose (**48**) were obtained in 9%, 2.7% and 6% yields with 6′-OH, 4-OH and 4′-OH free respectively (Fig. 13a).

The utilisation of this method in the synthesis of some biologically active oligosaccharides, agrocinopin A and B (see Section 5.) has been described by FRANZKOWIAK and THIEM (*68*). In addition to the compounds isolated by HOUGH, these authors also describe the preparation of the 6-hydroxy- and 1′-hydroxy-hepta-acetates in 24 and 11% yields respectively by essentially the same method. They also claim an increased yield of 9% for the 4-hydroxy compound.

Recently CAPEK et al. (*69*) have reported isolation of three penta-acetates by an adaption of this method where potassium carbonate was incorporated into the alumina. The penta-acetate fraction com-

References, pp. 175–184

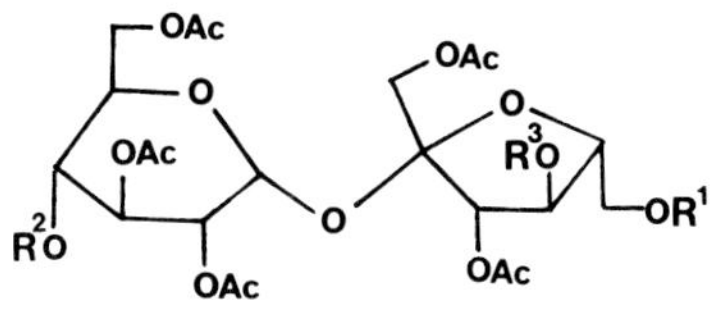

(46) R^1=H; R^2=R^3=Ac
(47) R^2=H; R^1=R^3=Ac
(48) R^3=H; R^2=R^2=Ac

a

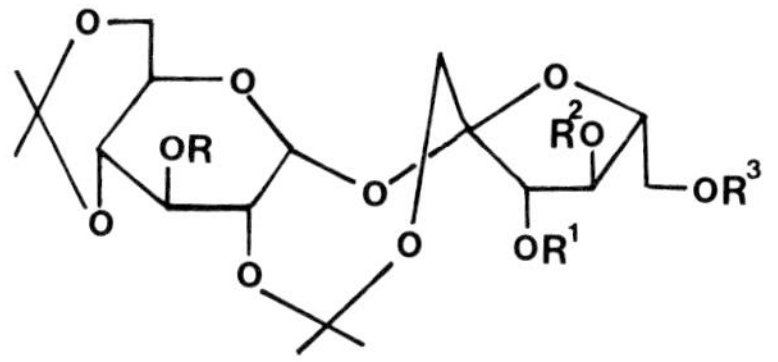

(34) R=R^1=R^2=R^3=Ac
(49) R=R^2=R^3=Ac; R^1=H
(50) R=R^1=R^3=Ac; R^2=H
(51) R=R^3=Ac; R^1=R^2=H
(52) R=Ac; R^1=R^2=R^3=H
(53) R=R^1=R^2=R^3=H

b

Fig. 13a, b. Selective de-esterification of sucrose esters

prised 24% of the reaction mixture which on elution with methanol yielded the 3',4',6'-trihydroxy-, 1',3',4'-trihydroxy- and 2,3',4'-trihydroxypenta-acetates in 9%, 67% and 24% yields respectively (70). Significantly de-esterification predominates on the fructofuranoside ring.

Selective de-O-acylation of 2,1':4,6-di-O-isopropylidenesucrose tetra-acetate (34) has been examined using methanolic ammonia at low temperatures (71). At −10 °C for 15 min, the reaction afforded an inseparable mixture of the 3'- and 4'-monohydroxy compounds (32%) (49) (50) and the 3',4'-dihydroxy compound (30%) (51) with starting material (24%) (Fig. 13b). When the reaction was repeated at −10 °C for 0.5 h followed by 3.5 h at 5 °C, the product distribution was as follows: 3',4'-dihydroxy compound (51) (35%); the 3',4',6'-trihydroxy compound (52) (44.5%), and the free acetal (53) (5.2%). From this (71), it was deduced that the order of hydrolysis of the four acetyl groups in (34) is: O-3' ≃ O-4' > O-6' > O-3. The low reactivity of AcO-3 was attributed to steric hindrance by the 4,6- and 1',2-O-acetal groups but the de-esterification is again observed to be selective for the ester groups on the furanoside ring. The 3',4'-dihydroxy derivative proved valuable in syntheses of the 3',4'-*lyxo*- and 3',4'-*ribo*-epoxides *via* the dimesylate (see Section 3.1.).

Novel approaches towards influencing the selectivity of certain hydroxyl groups towards acylation have utilised organometallic and inorganic reagents. Thus, AVELA and co-workers (*72*) have described the selective esterification of sucrose using transition metal chelates. It is presumed that complexation of the 'hard' ligating groups of a carbohydrate moiety with a "soft" metal cation will enhance the nucleophilicity of those groups involved and hence influence their selectivity towards acylating agents. Thus these authors claimed a 98% yield of monoesters when sucrose was treated with sodium hydride, $CoCl_2$, and acetic anhydride in the ratio 1:2:1:1.2 in DMF. G.c. and mass spectrometry indicated the reaction mixture to contain one major and two minor products; however no attempt at isolation or characterisation was made. Recently, the major products from the analogous reactions in pyridine for both acetylation and butyroylation have been isolated in 60% yield and have been identified by nmr as the 3'-*O*-esters (*73*).

2.3.1.2. Benzoates

An investigation into the relative reactivities of HO-3, HO-3', HO-4' and HO-6' towards benzoylation has been carried out by CLODE *et al.* (*74*). Thus treatment of 4,6:1',2-di-*O*-isopropylidenesucrose (**53**) with 3.3 molar equivalents of benzoyl chloride in pyridine-chloroform at 0 °C for 1 h afforded the 3',6'-dibenzoate (**54**), the 3',4',6'-tribenzoate (**55**) and the 3,3',6'-tribenzoate (**56**) in yields of 36%, 9% and 8% respectively (*75*) (Fig. 14a). The order of reactivity was therefore assigned to be HO-3' ≃ HO-6' > HO-4' > HO-3. The greater reactivity of the 3'-hydroxyl group was interpreted in terms of the *cis*-arrangements of HO-3' and the glucosidic oxygen O-2.

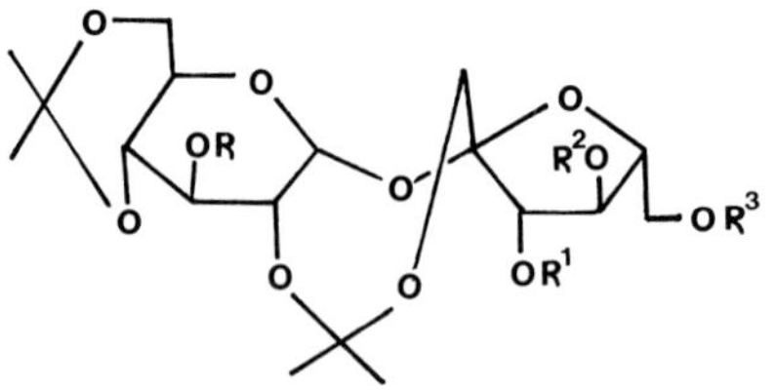

(**54**) R=R²=H; R¹=R³=Bz
(**55**) R=H; R¹=R²=R³=Bz
(**56**) R²=H; R=R¹=R³=Bz

Fig. 14a. Selective esterification of sucrose dicyclic acetal

A detailed study of the direct tribenzoylation of sucrose with benzoyl chloride in pyridine showed that the major product is the 6,1',6'-tri-O-benzoate and led to the conclusion that the order of acylation is 6-OH > 1'-OH : 6'-OH > 2-OH : 3'-OH, the reactivity at 1' and 3' being greater than expected (74). The taste properties of the products in relation to their chemical structures were investigated. It was observed that the tribenzoates are more bitter than the tetrabenzoates, the effect being increased by the presence of a 2- or 3'-benzoate group. 6,1',3'-Tri-O-benzoyl sucrose was synthesised by three routes: one approach used trimolar benzoylation of 6'-O-t-butyldiphenylsilyl sucrose and also gave 2,1',6'-tri-O-benzoylsucrose after chromatographic separation (76).

Treatment of sucrose with three equivalents of bis(tributylstannyl)-oxide and six molar equivalents of benzoyl chloride in toluene effected an enhancement of the nucleophilicity of primary and certain secondary hydroxyls with the formation of 2,3,6,1',6'-penta-O-benzoylsucrose in 87% yield (77).

2.3.1.3. Benzoylpropionates

A further classical method of hydroxyl group protection utilises the 3-benzoylpropionyl group (78, 79). Reaction of 2,3,4,1',3',4'-hexa-O-benzoylsucrose and 2,3,4,3',4'-penta-O-benzoylsucrose with 3-benzoylpropionic acid and N,N-dicyclohexylcarbodiimide in pyridine gives the respective 6,6'-di-O-(3-benzoylpropionyl) and 6,1',6'-tri-O-(3-benzoylpropionyl) derivatives. The benzoylpropionates are advantageous in their ease of preparation and subsequent facile removal due to their lability under mild conditions such as hydrazine in acetic acid and pyridine at room temperature where acetate and benzoate groups are retained.

2.3.1.4. Pivalates

Greater selectivity in esterification would be expected with sterically hindered acylating agents such as pivaloyl chloride (2,2-dimethylpropionyl chloride) and was studied by HOUGH et al. in 1978 (80).

Reaction of sucrose with 20 molar equivalents of pivaloyl chloride at $-40\,°C$ for 6 h followed by 24 h at ambient temperature in pyridine yielded 50% of the crystalline 4-hydroxy-hepta-O-pivaloylsucrose (81). This derivative provides a useful route to *galacto*sucrose since inversion at C-4 by mesylation followed by nucleophilic substitution with benzoate or acetate anions and de-esterification yielded the modified sucrose.

Slight alteration of the pivaloylation conditions of reaction led to the isolation of a series of compounds (57)–(62) ranging from the di-to the hepta-pivalate (Fig. 14b). These were separated chromatographically and then characterised by nmr and by derivatisation to the carbamates using trichloroacetyl isocyanate ($Cl_3CCONCO$). Treatment of sucrose with 3 molar equivalents of pivaloyl chloride yielded the 6,1',6'-tri-*O*-pivaloyl-(42%) and 6,6'-di-*O*-pivaloyl-(22%) derivatives.

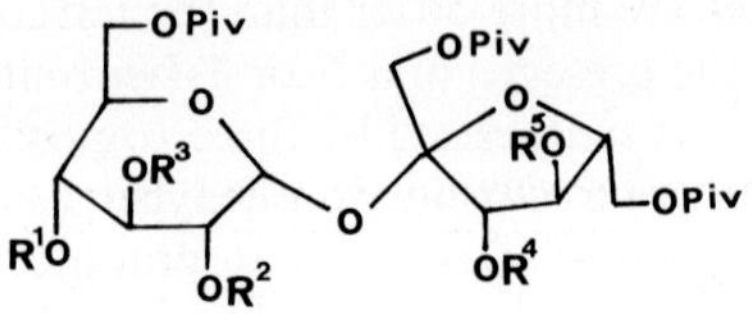

(57) R^1=OH; R^2=R^3=R^4=R^5=Piv
(58) R^1=R^2=OH; R^3=R^4=R^5=Piv
(59) R^3=R^4=OH; R^1=R^2=R^5=Piv
(60) R^1=R^4=OH; R^2=R^3=R^5=Piv
(61) R^1=R^2=R^5=OH; R^3=R^4=Piv
(62) R^1=R^3=R^4=OH; R^2=R^5=Piv

Fig. 14b

From the results hereby obtained, these authors (*81*) concluded that the 'reactivity profile' of sucrose towards pivaloyl chloride differs from that associated with esterification with other acid chlorides. This point was validated by comparison of the tetramolar tosylation (see Section 2.3.2.) and pivaloylation reactions. Whereas a reactivity order of O6 ≃ O6' > O-1' > O-2 had been established for the former reaction, in the latter the major product was the 6,1',4',6'-tetrapivalate and none of the expected 2,6,1',6'-tetrapivalate was detected. Two divergent routes to sucrose octapivalate via this reaction were subsequently suggested, each due to different reactivities of the partially pivalated derivatives towards further acylation;

a) 6,6'-OH > 1'-OH > 4'-OH > 2-OH > 4-OH > 3'-OH > 3-OH

b) 6,6'-OH > 1'-OH > 3'-OH > 3-OH > 4'-OH > 2-OH and 4-OH.

The array of compounds isolated in the course of this study (*81*) has been recognised as a potentially viable route to otherwise practically inaccessible protected sucrose esters of synthetic utility.

2.3.2. Sulphonates

There are two important aspects to the value of sucrose sulphonates as synthetic intermediates. Firstly, the esterifications of sucrose with limited quantities of various sulphonyl halides show selectivity, the general order of reactivity being 6-OH, 6'-OH > 1'-OH > 2-OH. Secondly, the resultant sulphonates then show significant differences in their preferences for substitution of the sulphonyloxy groups by nucleophiles, their reactivity varying according to their position in the conformation of the sucrose derivative, with 6-OR > 6'-OR > 4-OR > 1'-OR in terms of ease of their replacement.

Table 1. *Number of Isomers of the O-derivatives of Sucrose from Mono- to Octa-substituted*

Mono	8	Tri	56	Penta	56	Hepta	8
Di	28	Tetra	70	Hexa	28	Octa	1

In exploiting the subtle differences in reactivity of the eight hydroxyl groups of sucrose to produce partially substituted derivatives, it was appreciated that a large number of isomers could arise (Table 1). The higher reactivity of the primary hydroxyls does simplify these reactions to some extent. Thus treatment of sucrose with 3 molar equivalents of toluene-*p*-sulphonyl chloride (tosyl chloride) in pyridine was expected to yield the 6,1',6'-tritosylate (**63**) but chromatography revealed a mixture of products (Fig. 15) which on separation yielded a crystalline 6,6'-ditosylate (**64**) (*82, 83*), a mixed syrupy tritosylate fraction (*82, 84, 85*) and a small amount of the 2,6,1',6'-tetratosylate (*85*) (**65**). Further examination of the tritosylate fraction revealed the 6,1',6'-tritosylate as the major constituent but accompanied by an

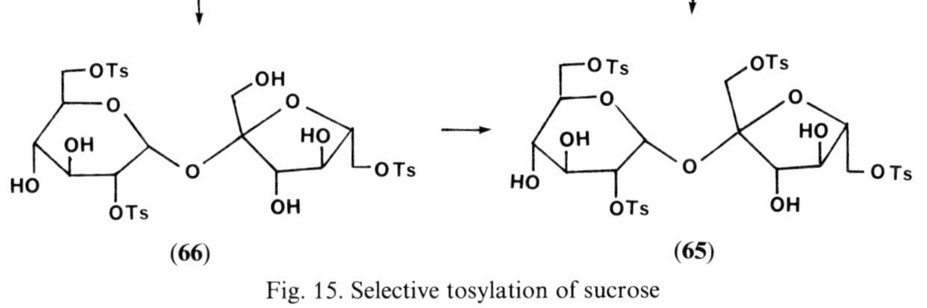

Fig. 15. Selective tosylation of sucrose

appreciable amount of the 2,6,6'-tritosylate (**66**). Acetylation of the tosylates yielded crystalline derivatives thus facilitating their purification. Tetramolar tosylation of sucrose gave the 6,1',6-tritosylate in 40% yield and the 2,6,1',6'-tetra-tosylate in 32% yield (*86*). The hindered 1'-OH is therefore less reactive than the other primary 6- and 6'-OH groups and the order of reactivity of sucrose to tosyl chloride is 6-OH, 6'-OH > 1'-OH > 2-OH > other OH groups.

Greater selectivity in the synthesis of the primary sulphonate esters of sucrose has been achieved by using the bulkier 2,4,6-trimethylbenzene ('mesitylene' or 'trimsyl') sulphonyl chloride (*87, 88*) or 2,4,6-tri-isopropylbenzene ('tripsyl') sulphonyl chloride (*89*). These reagents lead to derivatives which have the advantage of direct isolation, as the crystalline 6,1',6'-trisulphonates in > 50% yield without resorting to chromatography.

Sulphonate esters have been made available from studies in selective esterification (see Section 2.3.1.) and selective tritylation (see Section 2.1.1.). The 6,1',6'-tritosylate was synthesised (*90*) by tosylation of 2,3,4,3',4'-penta-*O*-acetylsucrose (**45**), followed by careful deacetylation. Similarly the re-arranged 2,3,6,3',4'-penta-acetate (**44**) affords the 4,1',6'-tritosylate (*91*). 1'-Mono-*O*-tosylsucrose is readily available by unimolar tosylation of the 6,6'-ditrityl ether (*92*). De-etherification of the hexa-acetate of 6,6'-di-*O*-tritylsucrose with aqueous acetic acid caused 4→6-acetyl migration, and the resultant hexa-acetate then gave the 4,6'-dimesylate (*93*). Unimolar tosylation of 6,1',6'-tri-*O*-tritylsucrose, in which the primary hydroxyls are all protected, gave the 2-tosylate in good yield (*94*). 6'-*O*-(*p*-Nitrobenzene sulphonyl) sucrose was obtained by de-tritylation and sulphonation of 6'-*O*-tritylsucrose hepta-acetate (**44**). Selective de-mono-*O*-acetylation of sucrose octa-acetate at C-6' affords a more convenient route to 6'-sulphonates (*67*). 4-*O*-Mesyl- and 4-*O*-tosyl-sucrose hexa-esters have been prepared either indirectly from 4,6-*O*-benzylidenesucrose hexa-acetate (*58*) or directly from sucrose 2,3,6,1',3',4',6'-heptapivalate (*81, 95*). Another route employed the reaction of the 2,3,6,3',4'-penta-acetate with tritylchloride which gave a mixture of the 2',6'-di-ether and 6'-ether. Subsequent mesylation, detritylation and acetylation of the separated products then yielded the corresponding 4-mesylate hepta-acetate and 1',4-dimesylate hexa-acetate (*93*).

Three hexapivalates of sucrose yielded dimesylates; that with free hydroxyls at C-3 and C-3' gave the 3,3'-dimesylate, whilst the one with free hydroxyls at C-2 and C-4 yielded the 2,4-dimesylate and that with free hydroxyls at C-3 and C-4' afforded, after chromatography, the 3-mesylate and 3,4'-dimesylate (*95*). Other partial mesylated products also isolated from the sucrose pivalates (see Section 2.3.1.4.)

included the 2,4,4'-trimesylate, 3,4,3'-trimesylate and 2,3,4,3',4'-pentamesylate (*95*).

Acetals of sucrose (see Section 2.2) provide another route to specific sulphonate esters. Thus 3,3',4',6'-tetra-*O*-acetylsucrose prepared from the 2,1';4,6-di-*O*-isopropylidene derivative undergoes selective trimsylation to give the 2,6,1'-trisulphonate (9%) and the 6,1'-disulphonate (52%) (*92*). Partial de-*O*-acetylation of the tetraacetate of the 2,1':4,6-di-*O*-isopropylidene-sucrose gave the 3,6'-diacetate and selective tosylation, followed by chromatography led to the isolation of the 3',4'-ditosylate, 4'-tosylate and 3'-tosylate in 4, 3 and 31% yields respectively (*96*). These results indicated that the order of reactivity of tosylation in the fructoside ring is 6'-OH > 1'-OH > 3'-OH > 4'-OH.

A useful route to 1',6'- and 1'-sulphonates utilises the selective substitution of the 6,1',6'-tri-*O*-trimsylate (**67**) with sodium benzoate in DMF (see Fig. 16) which progresses from the 6-*O*-benzoate (**68**) to the 6,6'-di-*O*-benzoate (**69**) (*97*).

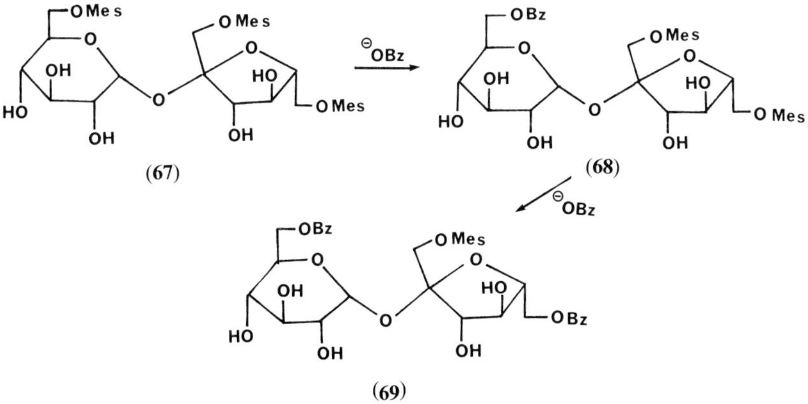

Fig. 16. Trimsylate derivatives of sucrose

The hexa-*O*-benzylsucrose obtained by the acid hydrolysis of 4,6-*O*-benzylidenesucrose was converted into the 4,6-dimesylate, a precursor of fluorosucroses (*98*). In addition, tritylation of the hexabenzyl ether, followed by *in situ* mesylation gave the 4-*O*-mesyl-6-*O*-trityl ether (*98*).

2.3.3. Chlorosulphates

HELFERICH and his co-workers (*99*) discovered an interesting multicentred reaction of sulphuryl chloride with carbohydrates whereby certain hydroxyl groups are replaced by chlorine atoms and other adjacent hydroxyls are transformed into cyclic sulphate esters. Thus they found that methyl D-glucopyranoside (**70**) when treated with sulphuryl chloride in pyridine gave rise to a methyl 4,6-dichloro-4,6-dideoxy-α-D-hexopyranoside 2,3-cyclic sulphate (**75**), later shown to be a galactopyranoside (*100*) (Fig. 17). J.K.N. Jones and his colleagues (*100, 101*) elucidated the stereochemical features of the reaction and demonstrated its versatility at different temperatures and pyridine concentrations, with the emergence of a series of simple but effective syntheses of chloro-deoxy derivatives.

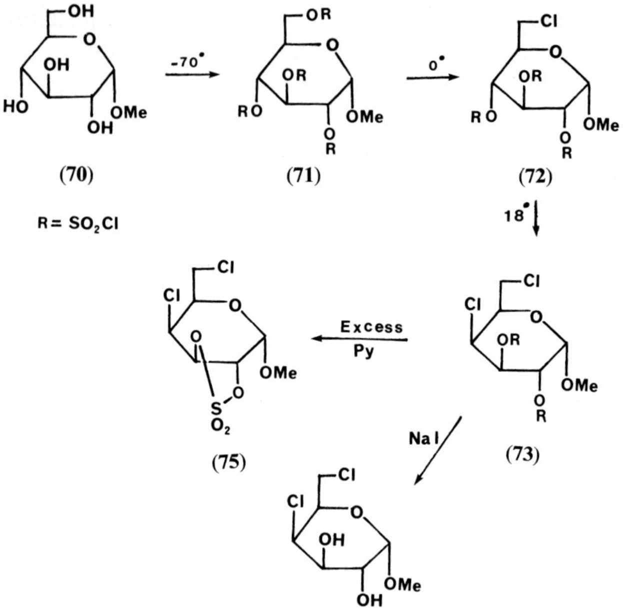

Fig. 17. Sulphuryl chloride reactions with methyl α-D-glucopyranoside

In general, reaction of a carbohydrate with sulphuryl chloride at −70° in the presence of minimal proportions of pyridine, and preferably diluted with chloroform, avoids cyclic sulphate ester formation (**75**), to convert each hydroxyl group to a chlorosulphuric ester. On raising the temperature of the same reaction to 0°, the primary chlorosulphate groups are converted to chloro substituents by nucleophilic substitution by the pyridinium chloride present. A further rise to room

References, pp. 175–184

temperature facilitates the replacement of chlorosulphate groups at secondary positions, essentially where the transition state is favoured, with the introduction of chloro groups but with inversion of configuration at that position. By these reactions methyl α-D-glucopyranoside (**70**) is first converted at $-70°$ to the 2,3,4,6-tetra(chlorosulphate) (**71**), thence at $0°$ to the 6-chloro-6-deoxy-2,3,4-tris(chlorosulphate) (**72**) and finally at room temperature to methyl 4,6-dichloro-4,6-dideoxy-α-D-galactopyranoside 2,3-bis(chlorosulphate) (**73**) (*101*). Steric and polar factors favour nucleophilic substitution at carbon 4 but not at carbons 2 and 3 (*102*). Chlorosulphuric ester groups are conveniently removed by treatment with methanolic sodium iodide to liberate the hydroxyl group with retention of configuration; thus the latter 2,3-bis(chlorosulphate) (**73**) yields methyl 4,6-dichloro-4,6-dideoxy-α-D-galactopyranoside (**74**).

$$I^- + Cl\!-\!SO_2\!-\!OR \rightarrow ICl + SO_2 + {}^-OR$$

Under similar reaction conditions followed by dechlorosulphation, sucrose (**5**) gave initially the 6'-monochloride (**76**) (43% optimum yield) followed by the 6,6'-dichloride (29%) (**77**) and then 4,6,6'-tri-chloro-(50%) (**78**) and 4,6,1',6'-tetrachloro (41%) (**79**) derivatives of *galacto*-sucrose (Fig. 18) (*103, 104*). The primary 1'-chlorosulphate resists reaction because of its hindered neo-pentyl nature whereas the 4-chlorosulphonoxy group readily undergoes substitution because steric and polar factors are more favourable than at carbon-1' and at all other secondary positions. The presence of a vicinal axial group and of a β-trans axial group inhibits substitution of a sulphonyloxy substituent on a pyranoid

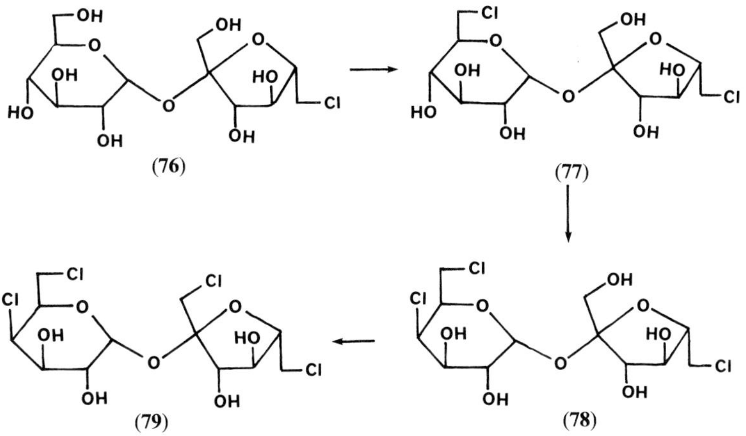

Fig. 18. Selective chlorination of sucrose

ring (*101, 102*). Thus the stereoselectivity observed in the substitution of chlorosulphonyloxy substituents in sucrose octa(chlorosulphate) proceeds in the sequence C-6′ > C-6 > C-4 > C-1′.

By omitting dechlorosulphation, PAROLIS (*105*) isolated crystalline 4,6,6′-trichloro-4,6,6′-trideoxy-*galacto*sucrose penta(chlorosulphate) in 50% yield after reaction of sucrose with sulphuryl chloride in pyridine, first at −78° and then at room temperature. When the sucrose-sulphuryl chloride-pyridine reaction is extended and completed at a high temperature (50–70°) further reactions occur at other positions giving the 4,6,1′,4′,6′-pentachloride of *galacto*sucrose (**80**) (*104*) but the 4′-chloro group is not introduced by direct nucleophilic substitution. Presumably it arises via the D-*lyxo*-3′,4′-epoxide (**81**) which then undergoes chloride anion attack exclusively at the 4′-position (Fig. 19). Under the original HELFERICH conditions for 48 h but at elevated temperatures (50°) sucrose reacted with sulphuryl chloride to give a complex mixture of products from which three complex derivatives were isolated by chromatography each containing 4,6-dichloro-4,6-dideoxy-2,3-sulpho-α-D-galactopyranosyl units (**75**) but with different substituents in the fructofuranosyl ring, namely a D-*lyxo*-3′,4′-epoxide, a 3′-ene and a 1′,4′,6′-trichloride (*106*).

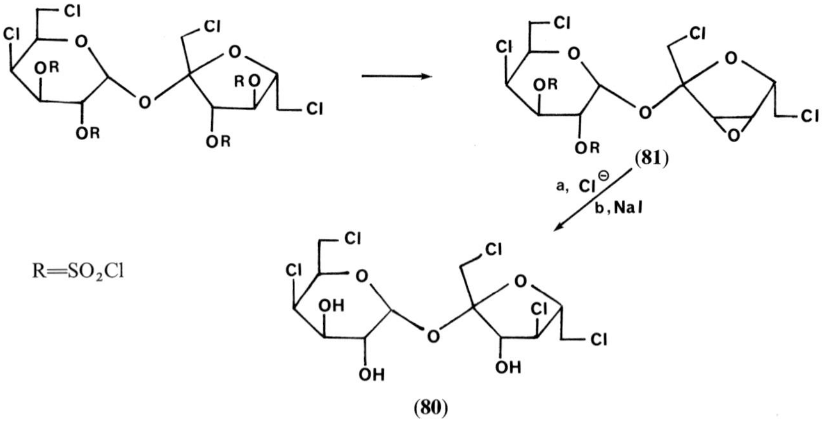

Fig. 19. Formation of a pentachloro derivative of sucrose

Ester substituents can be used as protective groups to exclude chlorination at specific positions during sucrose-sulphuryl chloride reactions (*12*). Thus the 6-acetate (**42**), or alternatively the 2,3,6,3′,4′-penta-acetate (**44**), yields the 4,1′,6′-trichloro derivative (**82**) (Fig. 20).

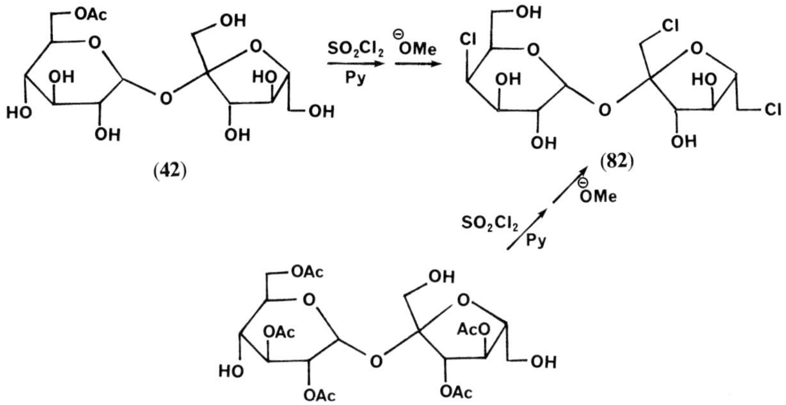

Fig. 20. Synthesis of 4,1',6'-trichloro-4,1',6'-trideoxy-*galacto*sucrose

The 6,6'-bis(chlorosulphate) and 6,1'6'-tris(chlorosulphate) (83) of sucrose were prepared (*107*) by treatment of the appropriate hexa- and pentabenzoates of sucrose respectively with sulphuryl chloride-pyridine-chloroform at −75°. Substitution of the 6,1',6'-tris(chlorosulphate) (83) with pyridinium chloride in either boiling chloroform or butanone gave the 6,6'-dichloride 1'-chlorosulphate (84) but in hexamethylphosphoric triamide (HMPT), an aprotic solvent, at 45° all three groups were displaced giving the 6,1',6'-trichloride (85) in 63% yield (Fig. 21). Using sodium chloride in HMPT for 24 h at 95° the yield

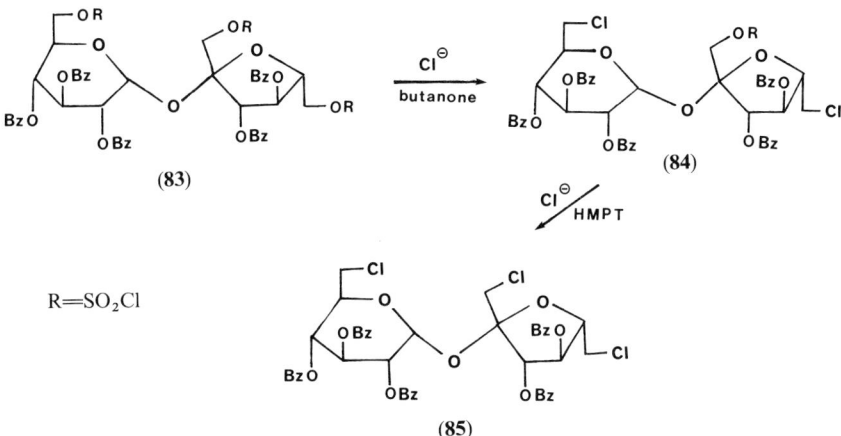

Fig. 21. Nucleophilic displacement of chlorosulphate groups

of the 6,1,6'-trichloride was increased to 88%. Likewise, 6,6'-dichloro-6,6'-dideoxysucrose 3,4,3',4'-tetra-acetate, prepared via the 1',2-isopropylidine derivative (**38**), afforded the 1',2-bis(chlorosulphate) (**86**) which reacted with pyridium chloride in DMF to yield after de-*O*-acetylation the 2,6,1',6'-tetrachloro-2,6,1',6'-tetradeoxy-*manno*sucrose (**87**) (Fig. 22) a compound with an exceedingly bitter taste (*107, 12*). This reaction is a rare example of nucleophilic substitution with inversion of configuration at carbon-2 of a pyranoside.

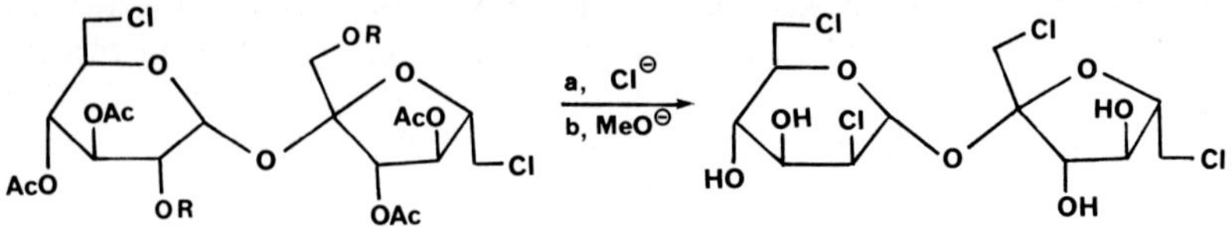

Fig. 22. Synthesis of 2,6,1',6'-tetrachloro-2,6,1',6'-tetradeoxy-*manno*sucrose

Attempts to use azide anion for the nucleophilic displacement of chlorosulphonyloxy substituents in sucrose to produce azidosucrose derivatives were unsuccessful. The use of other halides, such as iodide were also abortive (*12*). On the basis of this evidence, the S_N2 character of the nucleophilic substitution of chlorosulphoxyloxy substituents has been questioned, with the suggestion that the reaction is intramolecular of the S_Ni type that proceeds with inversion of configuration.

2.3.4. Other Esters

Enzyme mediated reactions offer considerable promise in the synthesis of specific sucrose esters. RIVA et al. (*108*) noted the regioselective esterification of sucrose with a protease (B subtilicin), with trichloroethylbutyrate in a solution of DMF, which by transferase activity gave sucrose 1'-butyrate.

Reaction of phthalic or succinic anhydrides with sucrose in DMF at 60 °C yields predominantly monosubstituted sucrose (*109*) mono(hydrogen phthalate) and mono(hydrogen succinate) respectively. On treatment with organometallic oxides these products yield organometallic derivatives of sucrose, such as the triphenyl-lead and tributylgermanium compounds (*109*). The organotin derivatives so prepared were shown to have biocidal properties.

Sucrose is converted by intermolecular esterification into insoluble carbonate polymer derivatives by reaction with phosgene in pyridine, ethylene dichloroformate in sodium hydroxide or diphenylcarbonate-sodium hydrogen carbonate (*110*). Partial esterification of sucrose with alkyl chloroformates at 0° in aqueous alkaline media afforded *O*-(alkoxycarbonyl)sucroses with a D.S. 4.9 to 8.0. A reaction at 0° with ethyl chloroformate in light petroleum to which 3 M sodium hydroxide containing sucrose was added, yielded an approximately trisubstituted *O*-ethoxycarbonylsucrose which precipitated from the reaction mixture in about 80% yield. This triethoxy carbonate polymerised on heating in the presence of a trace of sodium hydrogen carbonate at 150° *in vacuo* with elimination of diethylcarbonate and ethanol to give a thermosetting resin by the creation of a network of intermolecular carbonate ester linkages (*110, 111*). *O*-Alkyloxycarbonyl sucroses are remarkably resistant to acidic hydrolysis and are inert to invertase action. A crystalline octa-*O*-(ethoxycarbonyl)sucrose was prepared by treating the partially esterified derivative with ethyl chloroformate in pyridine (*111*).

Sucrose sulphates have been prepared, for example, by reaction with varying proportions of sulphur trioxide-DMF and pyridine to give products with degrees of substitution ranging from 1.5 to 6.8 (*112*). Their salts such as sucrose aluminium sulphate ("sucralfate") have been claimed to be effective in the treatment of gastric ulcers. An Xray crystal structure analysis (*113*) of potassium sucrose octasulphate heptahydrate revealed that the furanoside ring exists in a $^4T_5$ conformation, different from that found in simple sucrose derivatives.

Uncharacterised carboxymethyl ethers of sucrose have been prepared by treatment of sucrose with varying amounts of methyl bromoacetate in DMF in the presence of silver oxide. The pentacarboxymethyl sucrose was isolated in crystalline form and was converted into a uncharacterised trimethyl ester dilactone (*112*).

3. Derivatives

3.1. Anhydrides (Oxetanes) and Epoxides (Oxiranes)

Anhydro rings are readily obtained in sucrose, particularly from C-3 to C-6, C-3' to C-6' and C-1' to C-4'. Thus, 3,6:3',6'-dianhydrosucrose (**88**) is obtained (**83**) when the 6,6'-di-*O*-tosylate (**64**) is treated

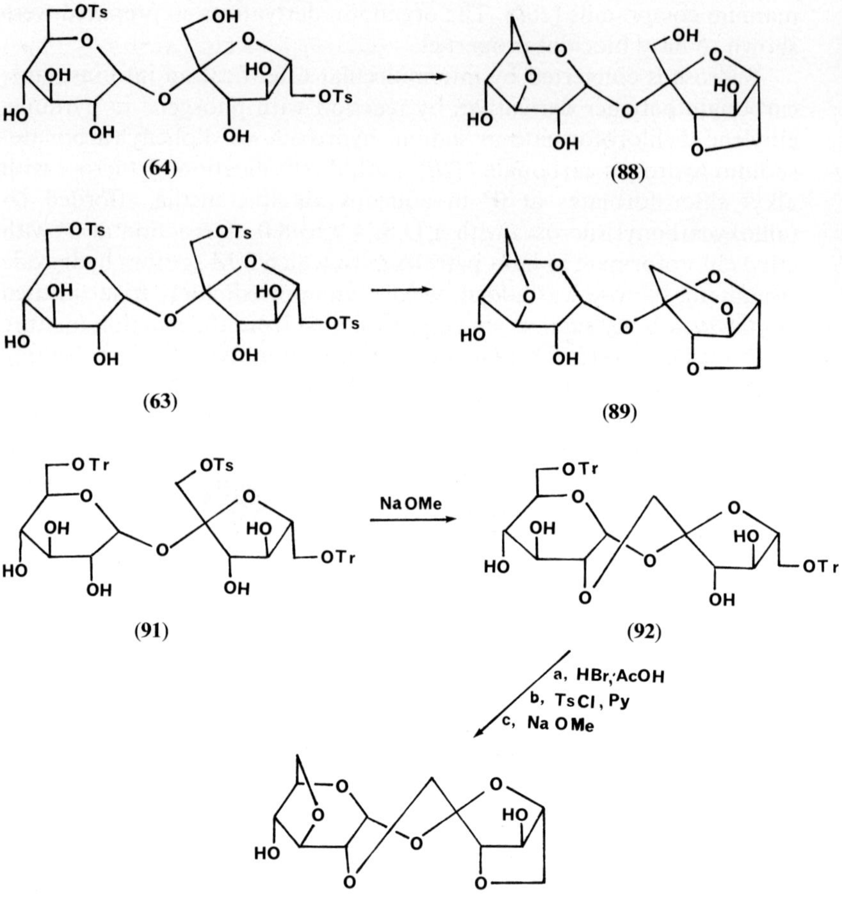

Fig. 23. Anhydro derivatives of sucrose

with methanolic sodium methoxide, when participation of the 3- and 3'-hydroxyl groups occurs in the intra-bimolecular substitution of the sulphoxyloxy substituents at O-6 and O-6' (Fig. 23). Similar reactions occur with the 6,1',6'-tri-O-tosylate (**63**) but with the additional formation of a 1',4'-anhydro ring the final product being 3,6:1',4':3',6'-trianhydride (**89**) (*90, 114, 85*) and not, as suggested on previous evidence, the 3,6:2,1':3',6'-trianhydride (**90**) (*82*). This trianhydride (**89**) has been isolated in two different crystalline forms (*85*). Treatment of either 2,6,1',6'-tetra-O-tosylsucrose (**65**), or 6,1',6'-tri-O-trimsyl-2-O-tosylsucrose with base followed a similar course and gave 3,6:1',4':3',6'-trianhydro-2-*O*-tosylsucrose in *ca.* 20% yield (*86*). In contrast to the behaviour of the 6,1',6'-tri-*O*-tosylate (**63**), the related 6,6'-di-*O*-tritylsucrose 1'-tosylate (**91**) was transformed by sodium methoxide into the bridged 2,1'-anhydride (**92**) which contains a dioxalane ring fused to a pyranoside of the strainless *cis*-decalin type (*92*). The difference in the pattern of cyclisation of the 6,1',6'-tri-*O*-tosylate (**63**) and the 1'-*O*-tosylate (**91**) has been attributed (*92*) to the prior formation of the 3',6'-anhydro ring in the former which moves the 4'-OH and the 1'-sulphonate groups into close proximity thus favouring the formation of the 1',4'-anhydro ring (**90**) rather than the 2,1'-anhydride (**92**). Subsequent de-tritylation, tosylation at C-6 and C-6' followed by anhydro formation with sodium methoxide-methanol, then gave 3,6:1',2:3,6-trianhydrosucrose (**90**) (Fig. 23).

The action of base on the 2-*O*-tosylate of the 6,1',6'-tri-*O*-tritylsucrose, wherein all of the primary hydroxyls are protected, yielded the expected 2,3-epoxy-D-*manno*-derivative in 40% yield together with the 3,4-*altro*-epoxide in 15% yield (*94*). The latter arose by migration of the epoxide ring by participation of the *trans*-3-hydroxyl group. The ratio of the two epoxides reflects their relative thermodynamic stabilities under equilibrating conditions.

Syntheses of 3,6:3',6'-dianhydro-sucrose (*83*), 1',4':3',6'-dianhydrosucrose (*43*) and 3',6'-mono-anhydro-sucrose (*44*) have been achieved by basic alcoholysis of 6,6'-di-*O*-tosyl-sucrose, 1',6'-di-*O*-mesyl-sucrose and 6'-*O*-(*p*-nitrobenzenesulphonyl)-sucrose respectively.

Halogeno-deoxy derivatives of sucrose can be used as alternative precursors of anhydrosucroses: thus 6-chloro-6-deoxysucrose heptaacetate with sodium methoxide in methanol yields 3,6-anhydrosucrose in 83% yield (*115*).

The base catalysed transformation of 4,1',6'-tri-*O*-tosylsucrose (**93**) into 3,6:1',4':3',6'-trianhydro-*galacto*sucrose (**94**), with inversion of configuration at C-4, has been attributed to the initial formation of a *galacto*-3,4-epoxide (**95**), followed by migration to the *gulo*-2,3-epoxide (**96**) and then attack by the 6-hydroxy at C-3 to give the 3,6-anhydro-D-galactopyranoside (**94**) (*116*) (Fig. 24).

Despite their potential synthetic importance, the preparation of epoxides (oxiranes) of sucrose has raised special difficulties because of the multifunctionality of the molecule and epoxide migration, which therefore requires careful control and the availability of a range of suitably blocked precursors. In 1973 the extensive reaction of sucrose with sulphuryl chloride led to the isolation of 4,6-dichloro-4,6-dideoxy-α-D-galactopyranosyl 3',4'-anhydro-1',6'-dichloro-1',6'-dideoxy-β-D-*ribo*-hexulofuranoside 2,3-sulphate in 17% yield (*106*). The synthesis

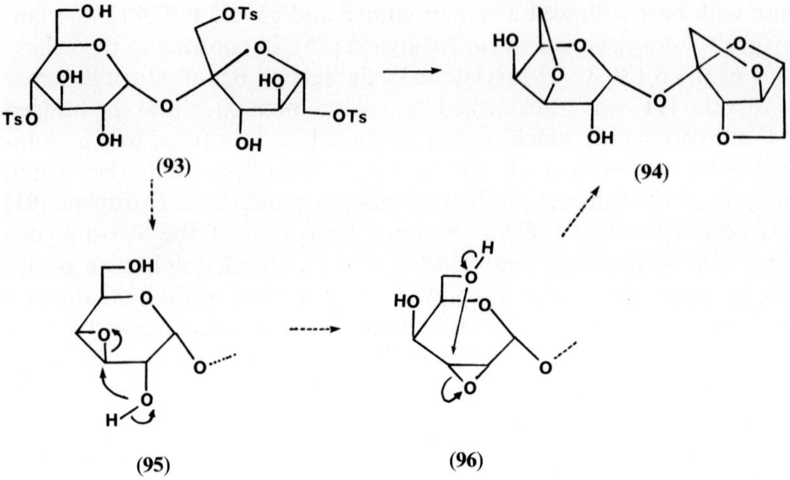

Fig. 24. A trianhydro derivative of *galacto*sucrose (**94**)

of suitably recently protected derivatives then enabled KHAN et al. (*96*) in 1978 to prepare 3'- and 4'-tosylates of sucrose which were then converted into the two 3',4'-epoxides, that is the D-*ribo*- (**98**) and D-*lyxo*- (**97**) hexulofuranosides.

● = OAc

A facile synthesis of a sucrose 3',4'-epoxide, namely α-D-glucopyranosyl 3,4-epoxy-β-D-*lyxo*hexulofuranoside (**97**) in 42% yield, was discovered by GUTHRIE et al. (*117*) by the direct reaction of sucrose in DMF with triphenylphosphine (TPP) and diethyl azodicarboxylate (DEAD) in the presence of acetic acid; the latter was incorporated with object of preventing 3,6- and 1',4'-anhydro ring formation. Derivatives of sucrose which protect the primary hydroxyl groups but leave the *trans*-vicinal 3',4'-hydroxyls exposed, such as 4,6:2,1'-di-*O*-isopropylidene sucrose (**53**), 2,3,6,1',6'-penta-*O*-benzoylsucrose and 6,1',6'-tri-*O*-tritylsucrose are all converted by DEAD-TPP into D-*lyxo*-3',4'-epoxides (**97**) in >80% yield (*12, 117*).

References, pp. 175–184

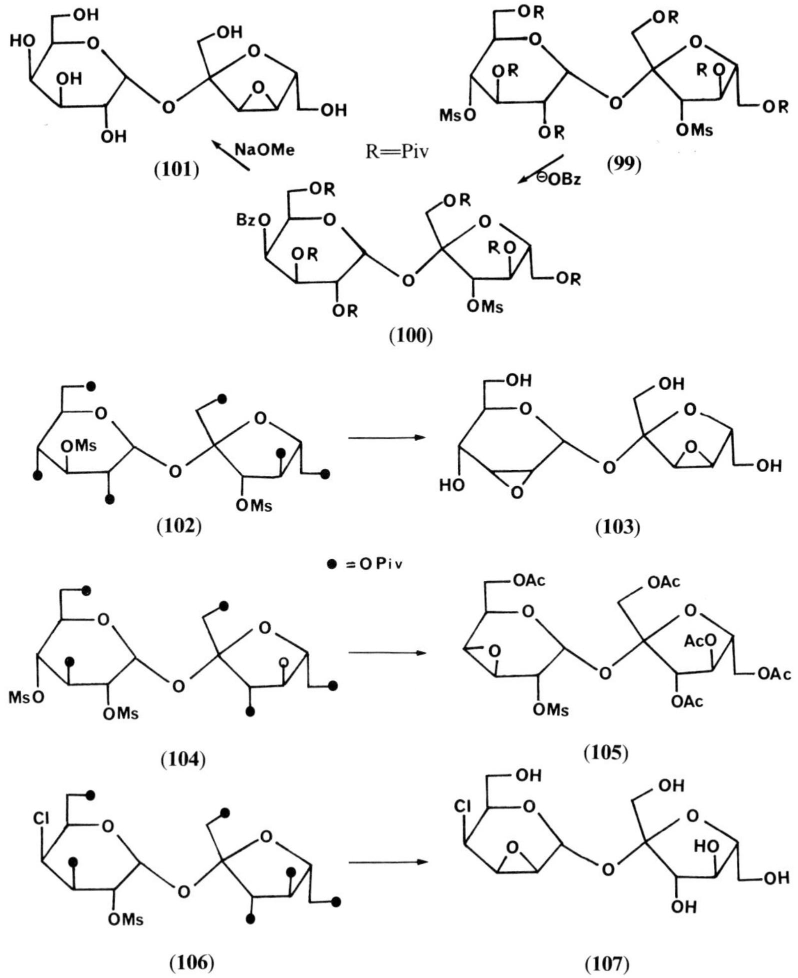

Fig. 25. Sucrose epoxides

Selected pivalate esters of sucrose (see Section 2.3.1.4.) have proved to be a useful source of epoxides (95) (Fig. 25). Thus one hexapivalate yielded the 4,3'-dimesylate (**99**) which underwent selective substitution with benzoate anion in DMF to give the 4-benzoyl-*galacto*sucrose 3'-mesylate (**100**). Subsequent treatment of this 3'-mesylate with base then afforded the D-*ribo*-3',4'-epoxide (**101**). Another hexapivalate gave the 3,3'-dimesylate (**102**) which with base yielded the 2,3:3',4'-diepoxide (**103**) in 37% yield. A third hexapivalate was converted into the 2,4-

dimesylate (**104**) which with sodium methoxide gave a complex mixture of products, but acetylation and column chromatography then gave the 3,4-epoxy-2-*O*-mesyl*galacto*sucrose penta-acetate (**105**) in 25% yield presumably as a consequence of the higher reactivity at C-4. The same 2,4-dimesylate (**104**) underwent selective substitution at C-4 with the nucleophilic chloride anion to give the 4-chloro*galacto*sucrose 2-mesylate (**106**) and subsequent cyclisation with base then gave the 4-chloro-2,3-epoxy-D-*talo*sucrose (**107**) in 67% yield (*95*) (Fig. 25).

The partially acetylated sucrose mixture obtained from sucrose octa-acetate by de-esterification with aluminium oxide and potassium carbonate (see Section 2.3.1.1.) yields after column chromatography three pentaacetates (*118*). Treatment of these with toluene-*p*-sulphonyl chloride in pyridine then afforded a 1'-tosylate, 1',3'-ditosylate 1',4'-ditosylate and 1',3',4'-tritosylate which on treatment with sodium methoxide were transformed into various 2,1'-anhydrides and 3',4'-epoxides (*118*).

3.2. Halides

Interest in deoxyhalogeno sucroses stems from their utility as intermediates in the synthesis of deoxyamino and thioderivatives although their potential in biochemistry and pharmacology, by analogy with the enhanced activity of antibiotic substances, has been recognised (*101*). Enhancement of the natural sweetness of sucrose by replacement of certain hydroxyls by chlorine atoms, coupled with the inhibition of invertase action, has stimulated much interest in the relationship between structure and the sweet response of deoxyhalogenosucroses (*103*).

6,6'-Dideoxy-6,6'-dihalogeno derivatives are readily obtained when 6,6'-di-*O*-tosylsucrose hexabenzoate (*119*), octa-*O*-mesylsucrose (*120*) and 6,1',6'-tri-*O*-tosylsucrose penta-acetate are treated, for example, with sodium iodide in butanone to yield 6,6'-dideoxy-6,6'-di-iodo derivatives. Under similar conditions 6,1',6'-tri-*O*-tripsylsucrose pentabenzoate gave the 6,6'-di-iodo compound in 98% yield (*89*). The relative reactivity of the primary positions of sucrose in S_N2 reactions involving sulphonyloxy displacements is therefore C-6 $\simeq$ C-6' > C-1'. Reaction of 6,6'-di-*O*-tosylsucrose hexa-acetate with sodium chloride in HMPT at 85 °C gave the 6,6'-dichloro derivative (51% yield) and 6-chloro-6-deoxy-6'-*O*-tosylsucrose hexa-acetate (34%) (*119*), indicating that substitution is more rapid at C-6 than C-6'. However the result was not duplicated in a similar experiment with octa-*O*-mesylsucrose which yielded an equimolar mixture of 6- and 6'-monochloro derivatives (*120*).

The 6,1',6'-trichloride (*108*) was synthesised from the 6,1',6'-tri-*O*-trimsylate (**67**) by substitution using lithium chloride in DMF containing a trace of iodine at 140° for 18 h (*88*) and as expected it could be then reacted with sulphuryl chloride-pyridine to give the 4,6,1',6'-tetrachloride of *galacto*sucrose (**79**) (Fig. 26). GUTHRIE and WATTERS (*121*) synthesised 1'-chloro-1'-deoxysucrose (**109**) from 6,1',6'-tri-*O*-trimsylsucrose (**67**) by initial substitution of the groups at C-6 and C-6' with benzoate anion in DMF, followed by replacement of the residual 1'-trimsyloxy group with chloride anion. Lithium chloride has a higher solubility in DMF and HMPT than sodium and potassium chloride and it is more effective in these nucleophilic substitutions. 1'-Chloro-1'-deoxysucrose was hydrolysed at a rate approximately 10 times slower than sucrose, due to the powerful inductive or field effect of the chloride atom. Furthermore it was not hydrolysed by invertase and was a competitive inhibitor of the enzyme, β-D-fructofuranosidase (*121*).

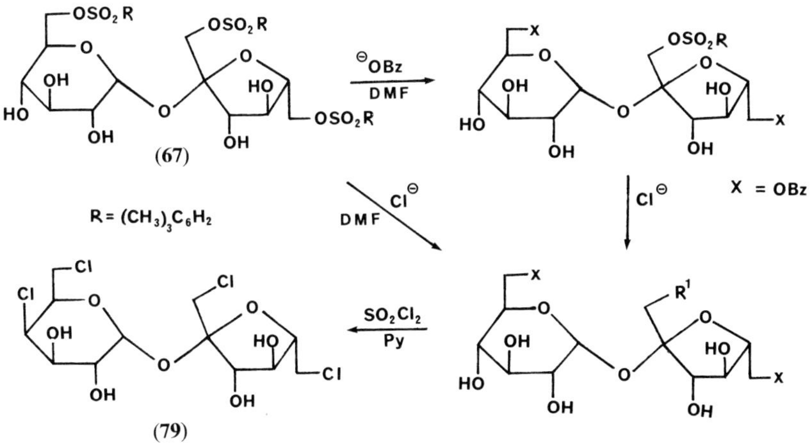

Fig. 26. Synthesis of chloro derivatives

Substitution at secondary positions is usually achieved in high boiling aprotic solvents of high dielectric constant but only where steric and polar factors favour the transition state for bimolecular nucleophilic substitution (*101*). Reaction of sucrose octamesylate with sodium bromide in HMPT at 85 °C afforded a mixture containing 4,6,6'-tribromo-4,6,6'-trideoxy-*galacto*sucrose pentamesylate (39%) and the 6,6'-dibromohexamesylate (36%), thereby revealing that the 4-position is more reactive than the 1'-position in sucrose and more favourable at C-4 for S_N2 reactions than any other secondary position in sucrose (*120*). Likewise, the 4,6'-dimesylate hexa-acetate with lithium bromide gave the 4,6'-dibromide of *galacto*sucrose and 6'-bromo-6'-deoxy-4-*O*-mesylsucrose (*93*); 4-chloro-4-deoxy-*galacto*sucrose was obtained from the 4-*O*-mesylsucrose hepta-acetate (*122*).

The direct replacement of selected hydroxyl groups in sucrose by chloride atoms via the chlorosulphates to give a range of mono- to pentachlorides has been achieved by the use of the sulphuryl chloride-pyridine reagent (*103*) (see Section 2.3.3.).

Direct replacement of hydroxyl groups by chlorine atoms can also be achieved by reaction of carbohydrates with methanesulphonyl chloride – DMF (*115*) and probably proceeds via the iminium ion $[(CH_3)_2N^+=CHOMs]Cl^-$ to give $[(CH_3)_2N^+=CHOR]Cl^-$ which then undergoes nucleophilic attack by chloride anion to give the chlorodeoxy product (*123*). Initially, sucrose yields 6,6'-dichloro-6,6'-dideoxysucrose (**77**) and then proceeds further to give a mixture of 6,1',6'-trichloro-6,1',6'-trideoxysucrose (**85**) together with the 4,6,6'-trichloride (**78**) and 4,6,1',6'-tetrachloride (**79**) of *galacto*sucrose, further examples of selective nucleophilic substitution at the secondary C-4 with inversion of configuration (*123*).

Using the protected derivative 2,3,1',3',4',6'-hexa-*O*-acetylsucrose, this reagent gave a mixture of the hexa-acetates of 4,6-dichloro-4,6-dideoxy-*galacto*sucrose (50%) and 6-chloro-6-deoxy-4-*O*-formylsucrose (40%) which were separable on silica gel columns (*115*). The formation of formate esters is a well known side reaction that accompanies the use of this reagent.

Selective reaction of sucrose with tris(dimethylamino)phosphine and potassium hexafluorophosphine in water gave a mixture of alkyloxy(dimethylamino)phosphonium salts which on treatment with sodium chloride in DMF gave 6-chloro-6-deoxysucrose and 6,6'-dichloro-6,6'-dideoxysucrose (**77**), isolated by column chromatography in 46% and 15% yields respectively (*123*). Similar results were observed with triphenylphosphine-*N*-halosuccinimide in DMF also, giving 6'- and 6-chloro and -bromo derivatives but in low yield (*124*).

An efficient method for the direct synthesis of the 6,6'-dichloride (**77**) and the 6,6'-dibromide of sucrose in high yields utilise either carbon tetrachloride (6 mol.) or carbon tetrabromide in the presence of triphenylphosphine and pyridine (*12, 125*). The high selectivity for the primary hydroxyls is attributed to the bulky complex of triphenylphosphine dihalide and pyridine.

References, pp. 175–184

Treatment of the 6,6'-dichloride (**77**) of sucrose with sodium methoxide converts it into the 3,6:3',6'-dianhydride (**88**) in high yield and in contrast the 4,6,6'-trichloride (**78**) of galactosucrose gave in two distinct stages, the 3,6-anhydro-4,6'-dichloride and then the 3,6:3',6'-dianhydro-4-chloride (*122*) due to the conformational transformation of an axial chloro group to an equatorial position.

Location of the position of substitution and determination of the stereochemistry of chloro-substituents introduced into the sucrose ring is readily determined from their ^{13}C-nmr spectra (*126*).

By using benzyl and trityl ether protecting groups, the 4-*O*-mesyl derivative of sucrose has been converted into 4-deoxy-4-fluoro-*galacto*sucrose by fluoride anion substitution (from tetrabutyl ammonium fluoride) in boiling acetonitrile for 3 days (*127*). The similarly protected 4,6-dimesylates of sucrose and *galacto*sucrose gave 4,6-dideoxy-4,6-difluoro-*galacto*sucrose and -sucrose derivatives respectively. By this approach 6-deoxy-6-fluorosucrose and 6-deoxy-6-fluoro-*galacto*sucrose were also synthesised (*98*). The possibility of fluoro-sucroses functioning as inhibitors of dextransucrase elaborated by *L. mesenteroides, Streptococcus mutans* and other bacteria is of interest in the prevention of tooth decay. The chlorosucroses significantly reduce acidogenesis from sucrose in the presence of *Streptococcus mutans in vitro* (*128*). Tetrachlorosucrose was found to have anti-acidogenic activity against plaque bacteria; furthermore it significantly reduces the ability of *Strep. mutans* to adhere to the tooth surface by preventing the formation of sticky extra cellular polysaccharides such as dextran (*128*).

3.3. Unsaturated, Deoxy and Branched-Chain Compounds

Terminal deoxy derivatives are readily available by either catalytic reduction of the appropriate primary halides or the exocyclic vinyl ethers. 6,1',6'-Trideoxysucrose has been synthesised by way of the 6,1',6'-trimesylate to the 6,1',6'-trichloride and finally hydrogenation with Raney nickel (*88*). The synthesis of 6,6'-dideoxysucrose utilised the reduction of the hexa-acetate of the 6,6'-dichloride with tri-n-butyltin hydride (*129*). Reductive dehalogenation of 4-chloro-4-deoxy-*galacto*sucrose by hydrogenation with Raney nickel in methanolic potassium hydroxide gave 4-deoxysucrose in 77% yield (*88*). Elimination of hydrogen halide from 6,6'-di-iodo derivatives of sucrose by means of silver fluoride in pyridine, destroyed the chirality at C-5 and C-5' to give the corresponding exocyclic vinyl ethers, namely 6-deoxy-β-D-*xylo*-hex-5-enopyranosyl 6-deoxy-β-D-*threo*-hex-5-enofuranosides (*83, 130*). The same approach was applied to 6-iodo- and 6'-iodosucrose esters to obtain β-D-fructofuranosyl 6-deoxy-α-D-*xylo*-hex-5-enopyranoside and α-D-glucopyranosyl 6-deoxy-β-D-*threo*-hex-5-enofuranoside respectively (*130*). Subsequent reduction of pyranoid exocyclic vinyl ethers can give both 6-deoxy-D- or L-hexosides. Hydrogenation of the 5-ene over palladium-on-charcoal gave 6-deoxysucrose and its L-*ido* isomer in yields of 10% and 45% respectively. On the other hand hydrogenation of the 5'-ene in ethylacetate-methanol with Pd/C afforded exclusively the 6'-deoxysucrose in 99% yield. The same product was also obtained by reductive dehalogenation of the 6'-iodide using hydrazine hydrate in the presence of Raney nickel (*130*).

The 6- and 6'-*C*-methylsucrose have been obtained by oxidation of the appropriate heptabenzyl ethers with the Pfitzner-Moffatt reagent followed by alkylation with methyl magnesium iodide and then removal of the protecting groups (*131*). Likewise oxidation of a hepta-*O*-pivaloylsucrose with methyl sulphoxide-acetic anhydride gave the 4-ulose which reacted with methylmagnesium iodide to give two isomeric 4-*C*-methyl derivatives, the *gluco* and *galacto*isomers, each assigned from ^{13}C-nmr spectra (*132*). Reaction of the 4-ulose with allyl magnesium bromide was more stereospecific yielding predominantly the *galacto*-isomer with the 4-allyl substituent equatorial (*132*).

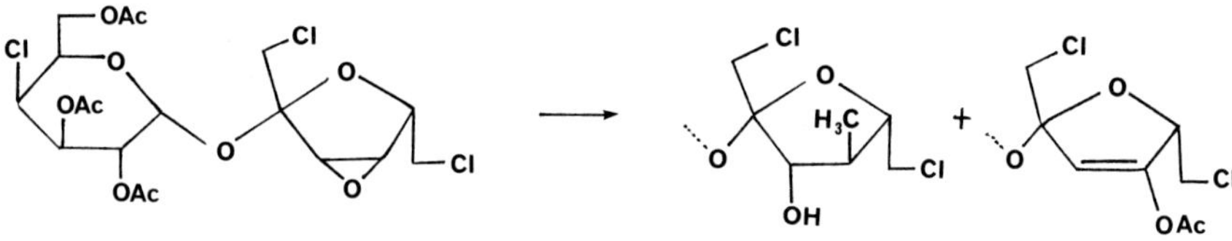

Fig. 27. A *C*-methyl derivative of sucrose

A novel approach to the synthesis of a 4'-*C*-methyl derivative (*111*) involved ring opening of the 3',4'-*lyxo*-epoxide (**110**) with lithium dimethylcuprate, a carbon nucleophile, which gave the required product in 22% yield accompanied by the 3'-eno derivative (**112**) in 16% yields (*133*) (Fig. 27). An entry to 1'-*C*-substituents was obtained by oxidation of HO-1' in the 6,6'-bis(*tert*-butyldiphenylsilyl)sucrose pentaacetate with dimethyl sulphoxide and trifluoroacetic anhydride to give the 1'-aldehyde derivative (**113**). This aldehyde was then reacted with the stabilised Wittig reagent, $Ph_3P=CHCO_2Et$, to give the 1'-ethoxycarboxylmethylenederivative (**114**) (Fig. 28). Deprotection with methanolic sodium methoxide was accompanied by Michael addition to yield 2,1'-anhydro-1'-methoxycarbonylmethylsucrose (*134*) (**115**).

References, pp. 175–184

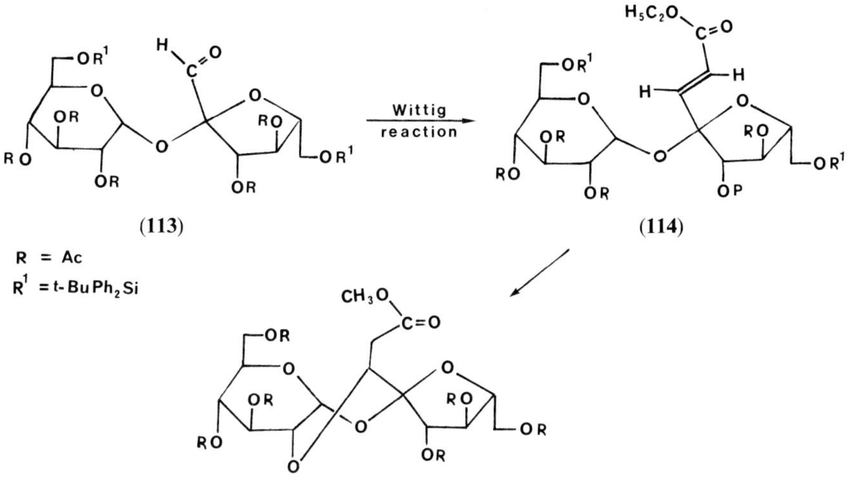

Fig. 28. Wittig reaction of sucrose 1'-aldehyde

3.4. Nitrogen-Containing Compounds: Azides, Amines and Morpholines

Azidodeoxy derivatives are the intermediates of choice for the synthesis of aminodeoxy compounds by catalytic reduction. The aminodeoxy sugars are of interest because of their potential biological activity since they occur as components of antibiotics, bacterial polysaccharides and glycoproteins. The azides are prepared either by nucleophilic ring opening of epoxides or substitution of sulphonate and halide derivatives of sucrose using sodium or potassium azide ion in a suitable solvent (*121*).

In a general study of the relative reactivities of the groups in octa-*O*-mesylsucrose (**116**), sodium azide in butanone-DMF (10:1 v/v) at 110 °C gave the 6,6'-diazide (**117**) in 75% yield but under more forcing conditions in butanone-DMF (1:2 v/v) at 130 °C a triazide was isolated in 20% yield in addition to the diazide and characterised as 4,6,6'-triazido-4,6,6'-trideoxy-*galacto*sucrose (**118**) (Fig. 29). In HMPT at 90 °C the yield of the triazide was increased to 60% and the reaction progressed to form the 4,6,1',6'-tetra-azide (*119*), isolated in 10% yield after column chromatography (*120*). 6,6'-Diamino-6,6'-dideoxysucrose was prepared (*135*) from 6,6'-di-*O*-tosylsucrose hexabenzoate by sequential nucleophilic substitution with azide anion in DMF, de-esterification and then hydrogenation over Pd/C. Repetition of this reaction sequence with the 4,1',6'-trimesylate gave 4,1',6'-triamino-4,1',6'-tri-

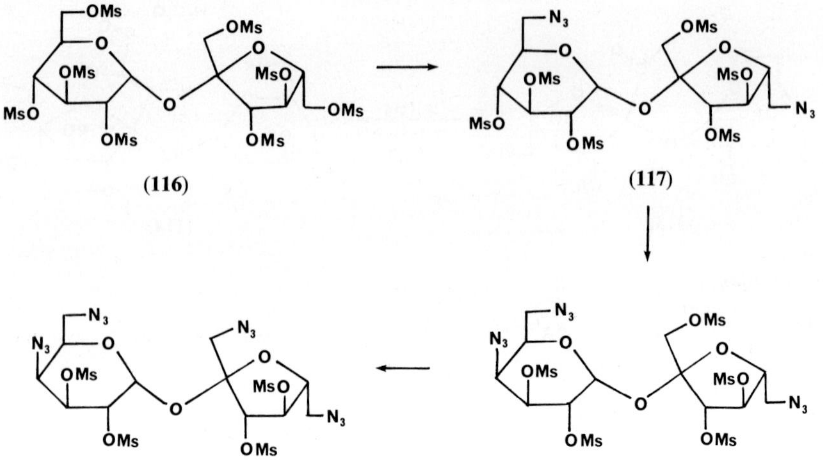

Fig. 29. Selective substitution of mesyl groups with azide

deoxy-*galacto*sucrose and the 6,1',6'-tritosylate was similarly converted to the 6,1',6'-triamine (*136, 137*). In general the azido group is difficult to reduce catalytically in the presence of *O*-esters. 6'-Amino-6'-deoxy- and 1',6'-diamino-1',6'-dideoxysucrose have been prepared by the same route, starting from the appropriate trityl ethers (*138*). 4-Amino-4-deoxy*galacto*sucrose was similarly prepared from a hepta-*O*-pivaloyl sucrose (*95*).

A selective reaction of 6,1',6'-tri-*O*-tripsylsucrose with sodium azide in HMPT was observed at 50° giving the 6-azide in 88% yield, due to the higher reactivity at the 6-position than the other primary groups; at 85° the 6,6'-diazide was obtained in 84% yield (*89*). In summary, the reactivity of the sulphonyloxy substitution in sucrose sulphonates by azide proceeds in the order: C-6 > C-6' > C-4 > C-1' and can be exploited by careful control of the temperature and solvent composition of the reaction.

N-Acylation of the 6,1',6'-triamine gave the corresponding 6,1',6'-triacetamide with acetic anhydride in methanol but using pyridine solvent both *N*- and *O*-acetylation occurs forming 6,1',6'-triacetamido-6,1',6'-trideoxysucrose penta-acetate (*137*).

Ring opening of the 3',4'-*ribo*-epoxide (**98**) with sodium azide in aqueous ethanol, in the presence of ammonium chloride, at 80° gave only the 4'-azido-4'-deoxy-β-D-*xylo*hexulofuranoside (**120**) (Fig. 30). Similar stereo-selectivity for attack at C-4' was observed when the 3',4'-*lyxo*-epoxide (**97**) was treated in the same manner with the sole forma-

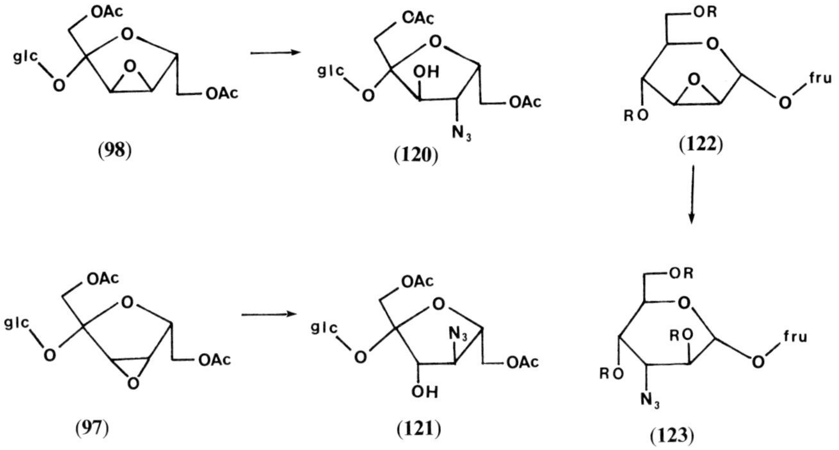

Fig. 30. Synthesis of azido sucroses

tion of 4'-azido-4'-deoxysucrose (121), as governed by steric and polar factors in the transition state (11, 139). For the same reasons, ring opening of the 2,3-epoxide (122) occurred by axial attack at C-3 by azide to give 3-azido-3-deoxy-α-D-altropyranosyl β-D-fructopyranoside (123) (94).

Morpholino-glucopyranosides (124) have been synthesised from sucrose by selective oxidation (140) with lead tetra-acetate of the fructofuranoside ring to a dialdehyde (125) followed by reductive amination with sodium borohydride in the presence of an amine such as benzylamine, methylamine and 1-amino-2,3-dihydroxy-propane (141). Some of these morpholine derivatives were sweet ($R = COCH_3$) and others bitter ($R = CH_2Ph$) (Fig. 31).

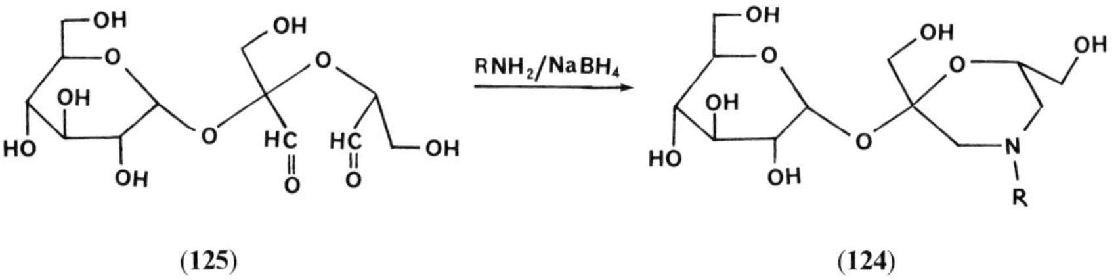

Fig. 31. Synthesis of morpholino glucopyranosides

The above dialdehyde (125) undergoes Fischer cyclisation with both nitromethane and nitroethane in the presence of methanolic sodium methoxide (142) (Fig. 32a). The former gave a mixture of four diaster-

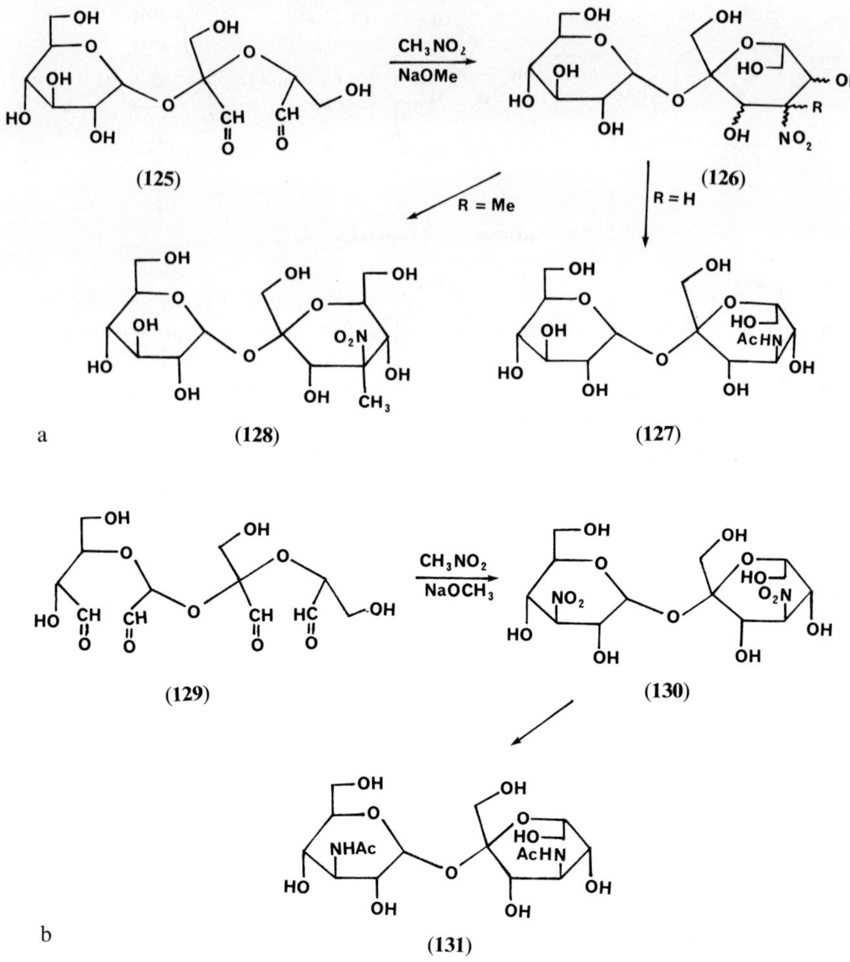

Fig. 32a, b. Condensations of nitroalkanes with dialdehydes derived from sucrose

eoisomeric nitropyranosides (126) from which the major component was isolated by chromatographic fractionation. Rapid hydrogenation using Raney nickel catalyst followed by N-acylation afforded α-D-glucopyranosyl 4-acetamido-4-deoxy-β-D-glucoheptulopyranoside (127). An unexpected epimerisation at C-5′ was encountered with nitroethane, since the only product encountered was α-D-glucopyranosyl 4-deoxy-4-C-methyl-4-nitro-α-L-glucoheptulopyranoside (128).

Nitromethane condensation of the tetra-aldehyde (129), produced by periodate oxidation of sucrose, gave 3-nitro-3-deoxy-α-D-glucopyranosyl 4-nitro-4-deoxy-β-D-heptulopyranoside (**130**) as the major product (*142*). Catalytic hydrogenation of the dinitro compound (**130**) and *N*-acetylation then afforded the new disaccharide, 3-acetamido-3-deoxy-α-D-glucopyranosyl 4-acetamido-4-deoxy-β-D-*gluco*heptulopyranoside (**131**).

3.5. Sulphur Derivatives (*143*)

Substitution of the bromo groups in the readily available 6,6'-dibromo-6,6'-dideoxysucrose hexaacetate, thioacetate and *N,N*-dimethyldithiocarbamate gave the corresponding derivatives of 6,6'-dithiosucrose. However when the dibromide was reacted with thiourea to give bis-thiouronium salt, sequential decomposition with sodium metabisulphite gave the bridged 6,6'-epidisulphide (**133**), isolated as the hexa-*O*-acetate in 19% overall yield. This disulphide (**133**) arose from the 6,6-dithiol (**132**) by aerial oxidation, a facile reaction observed with other sugar dithiols. The 11-membered trioxadithiaundecane ring system can exist in two conformations with S-6/S-6' being up/down (**133a**) or down/up (**133b**), and ^{1}H-nmr studies revealed the existence of an equilibrium between both conformers (Fig. 33). Desulphurisation of the

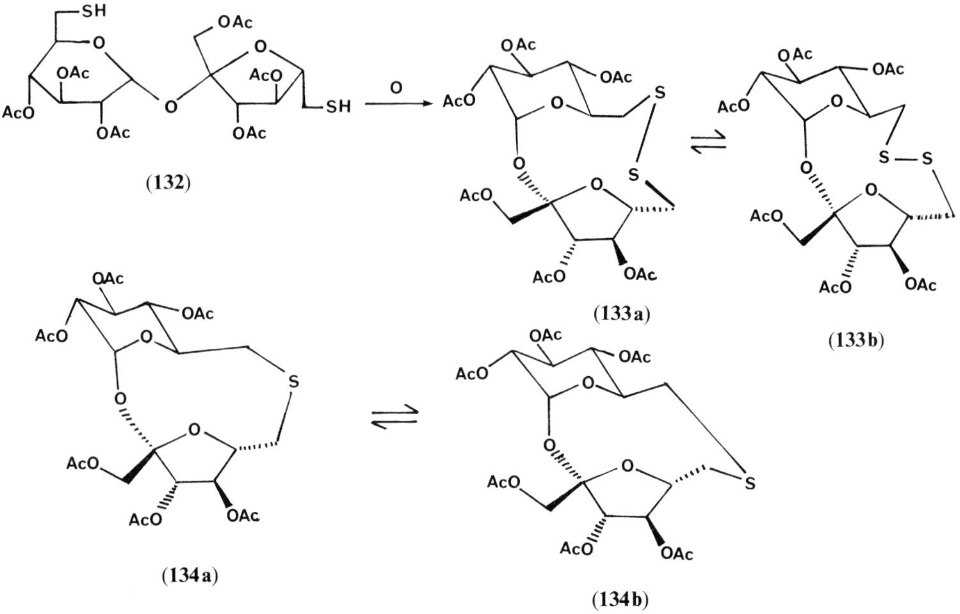

Fig. 33. 6,6'-Episulphide and 6,6'-epidisulphide derivatives of sucrose

6,6'-dithio derivatives with Raney nickel each gave 6,6-dideoxysucrose (*143*).

When the hexaacetate of the 6,6'-dibromide was treated with potassium *O*-ethyldithiocarbonate (EtOCS$_2$K) in DMF, a complex reaction sequence occurred leading to the monosulphide (**134**) namely 6,6'-epithiosucrose hexa-acetate (**134**) (Fig. 33). Alternatively the reaction of the dibromo derivative with potassium trithiocarbonate (K$_2$CS$_3$) gave a mixture from which the 6,6'-episulphide (**134**) was isolated in 24% yield and the 6,6'-epidithiosucrose (**133**) in 7% yield.

The molecular conformation of the episulphide, as revealed by X-ray crystallography showed that the pyranose ring has the usual $^4C_1$ conformation but that the fructofuranoside ring exists in a $^4T_5$ conformation, different from that in sucrose. As with the epidisulphide (**133**), the broadening of the nmr resonance of the epithiosucrose (**134**) arose from conformational instability of the trioxathiadecane ring. Due to interconversion of the two conformers (**134a** and **134b**) at elevated temperatures sharpening of the resonances occurred (*143*).

Oxidation of the episulphide with sodium metaperiodate afforded exclusively the (*R*)-sulphoxide and further oxidation with hydrogen peroxide then gave the sulphone.

Opening of the manno-2,3-epoxide (**135**) with potassium thioacetate and ammonium chloride in aqueous ethanol occurred by axial attack at C-3 to give 3-*S*-acetyl-3-thio-altropyranoside (**136**) (Fig. 34). Treatment of this epoxide (**135**) with ammonium thiocyanate in 2-methoxyethanol resulted in a complex mixture of products from which the

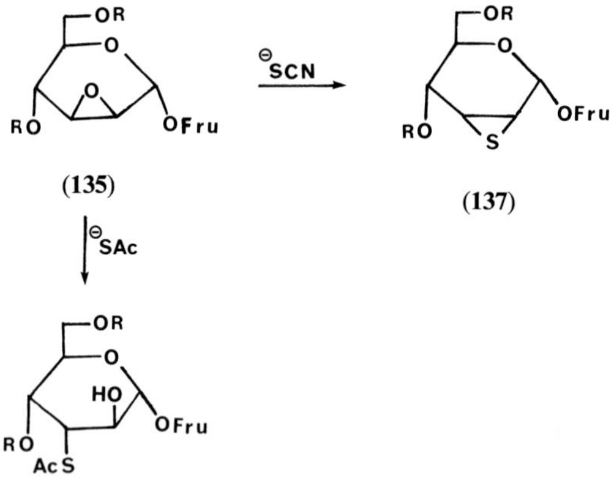

Fig. 34. Sulphur derivatives from sucrose 2,3-epoxide

References, pp. 175–184

major product was isolated in 23% yield (*94*). It was not a thiocyanate and arose by the known conversion of epoxides to episulphides of opposite configuration, in this case the 2,3-*allo*-episulphide (**137**).

4. Enhancement of Sweetness: Structure Activity Relationships

There is compelling, albeit circumstantial, evidence that all sweet organic compounds have a common feature within their diverse chemical structures. This feature consists of two electronegative atoms, designated **A** and **B**, which are close but separated by 2.5–4.0 Å and with a hydrogen atom covalently attached to one of these atoms giving **AH**. SHALLENBERGER and ACREE (*144*) postulated that the **AH/B** unit interacted with a similar structural feature on the peptide parts of the taste bud protein, such as a hydroxyl (OH) or imino(NH) group, acting as the **AH** and a carbonyl (C=O) as the **B** unit to give a pair of intermolecular hydrogen bonds which then initiate the sweetness sensation (Fig. 35). Sugars are sweet because of the presence of at least two hydroxyl groups, one acting as the **AH** and the oxygen atom of the other as the **B** unit (Fig. 35) (*145*). On the sweetness scale, where sucrose (1 ×) is the standard reference, the carbohydrates are not very

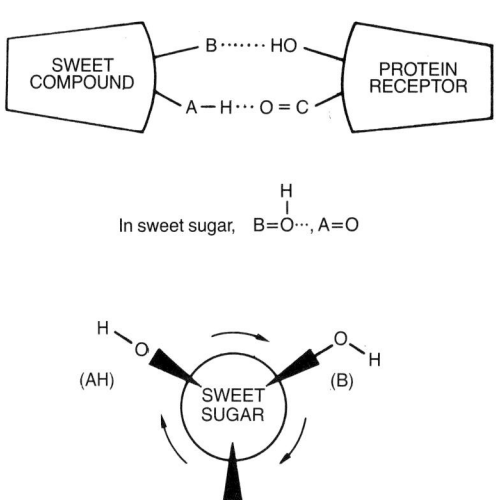

Fig. 35. Clockwise AH/B/X glucophore suggested for sucrose

sweet because of their hydrophilic nature whereas the sweeteners of high intensity, such as saccharin (400–500 ×) and aspartame (100–200 ×) are more strongly bound to the taste buds because they are more lipophilic. KIER (*146*) recognised the importance of lipophilicity in sweetness enhancement and introduced an additional hydrophobic factor **X** which acts in harmony with the **AH/B** unit to lock or guide the sweet compound onto the receptor site. In addition, when the glucophore resides in a chiral compound such as a sugar or an amino acid, the **AH/B/X** triad requires a clockwise arrangement of these groups respectively when viewed from the position of the proteinaceous receptor site (Fig. 35) (*147*).

The **AH/B** of sucrose was assigned (*147*) to the 2-hydroxyl of the glucosyl unit and either the 1′- or the 3′-hydroxyl of the fructosyl unit (**138**) from a consideration of the proximity of hydroxyls in the conformation of sucrose, coupled with the knowledge that methyl β-D-fructofuranoside is tasteless and methyl α-D-glucopyranoside is only slightly sweet (0.1 ×). Hence the sweetness of sucrose is due to the combined action of two hydroxyls, one from the glucosyl unit and the other from the fructosyl unit.

The sweetness of L-sucrose and other L-sugars such as L-glucose and L-fructose has been attributed to an interchange of the role of the two hydroxyls from **AH/B** in the natural D-form to **B/AH** in L-form (*147*). This interchange is not possible in other sweet chiral compounds such as the amino acids where D-forms are usually sweet whereas the L-forms are bitter (*146*).

One lipophilic centre **X** was located at the axial 4-position of sucrose because of the lack of sweetness of *galacto*sucrose, the sweetness of 4-deoxysucrose (1 ×) and the enhanced sweetness of 4-chloro-4-deoxy-*galacto*sucrose (5 ×). The greater sweetness of 6′-chloro (20 ×) and 1′-chloro (20 ×) derivatives suggested alternative lipophilic centres (**X**) in sucrose and it is significant that the three **X** locations are on the upper face of the molecule. On the other hand 6-chloro-6-deoxysucrose is not sweet but bitter and so are the 6-benzyl ether, the 6-*O*-benzoate and the 6-phosphate, whilst the 6-*O*-acetate is slightly sweet, and 6-deoxysucrose and 6-*O*-methylsucrose are as sweet as sucrose. Thus the size of the substituents at C-6 is critical with any significant increase resulting in the loss of sweetness (*147, 148*) which is also observed in 6-*O*-(α-D-galactopyranosyl)sucrose (raffinose) which is only 0.2 times as sweet as sucrose. The sweetness of the 1′-deoxy derivative which lacks the 1′-hydroxyl clearly favours the triangular glucophore in sucrose at the axial 4-position (**X**), 2-hydroxyl (**B**) and 3′-hydroxyl (**AH**) (**138**). This assignment is supported by the lack of sweetness of 3′-*O*-acetylsucrose where the **AH** is blocked by the acyl group (*73*).

References, pp. 175–184

(138)

Increasing the lipophilicity of sucrose by combination of two or more chloro groups at C-4, C-1' and C-6' had a synergistic effect resulting in greater sweetness, such that 1',6'-dichloride was 70 times and the 1',4-dichloride 120 times sweeter than the parent sucrose. On the other hand the 6,6'-dichloride was not sweet due to the adverse effect of the 6-chloro substituent. A similar effect was observed in the 6,1',6'-trichloride which was only 25 times sweeter than sucrose (147).

The synergistic effect of a combination of the chloro groups at C-4, C-1' and C-6' in 4,1',6'-trichloro-4,1',6'-tridoexy-*galacto*sucrose ("sucralose") (82) raised the sweetness to 650×, with a similar taste profile to sucrose and no bitter aftertaste (149) (150). Since sucralose was non-toxic non-cariogenic, not metabolised and hence non-nutritive and 60 times more stable to hydrolysis by acid than sucrose, it was selected by Tate and Lyle Plc and Johnson and Johnson (U.S.A.) for commercial development as a high intensity sweetener.

Additional chloro groups in sucralose led to either a further enhancement or reduction in sweetness depending upon the location of the new substituent. Thus the 4,1',4',6'-tetrachloro derivative of *galacto*-sucrose (139) was 2,200 times sweeter (104), a four fold increase in the sweetness of sucralose due to replacement of the 4'-hydroxyl by a chloro group on the upper face of the molecule, with each **AH**, **B**, **X** in a clockwise direction when viewed from the point of contact with the sensory protein (147) (148).

The *sorbo*-tetrachloride (140) with a configuration at C-3' and C-4' opposite to that of the *fructo*-isomer was found to be only 200 times sweeter (145). Furthermore replacement of the equatorial 2-hydroxyl in the 6,1',6'-trichloride by an axial chloro group to give the 2,6,1',6'-*manno*-tetrachloride (87) completely obliterated the sweetness rendering the molecule as bitter as quinine (151). An equatorial 2-hydroxyl is an essential requirement for sweetness in sucrose and its derivatives.

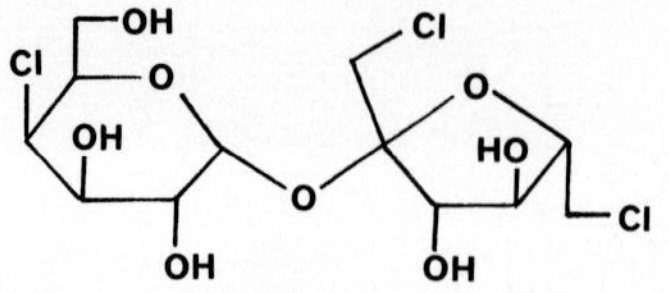

Fig. 36. Sweet and bitter chloro derivatives prepared from sucrose and trehalose

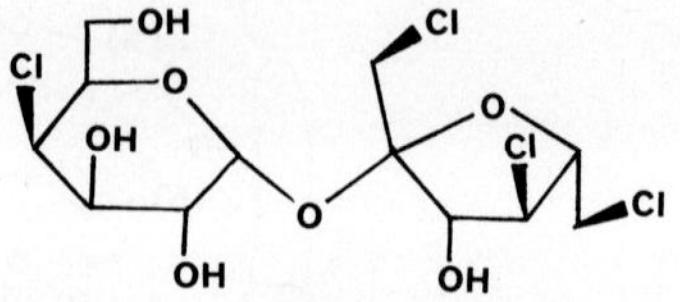

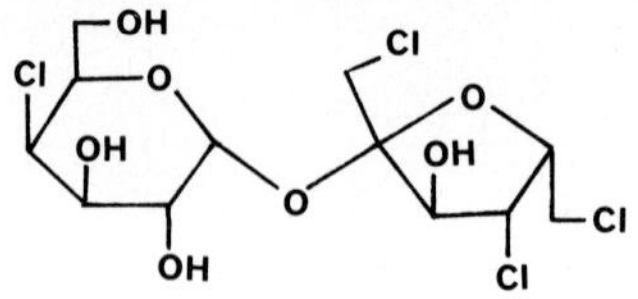

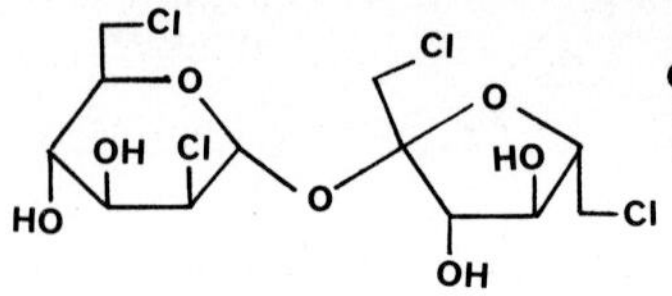

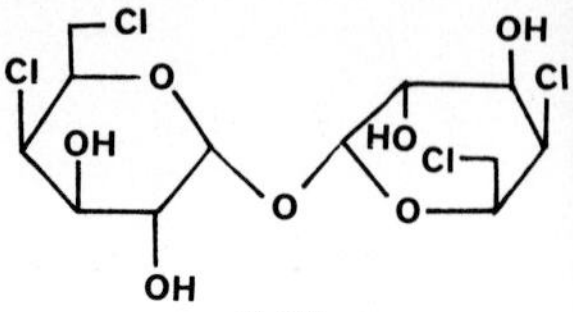

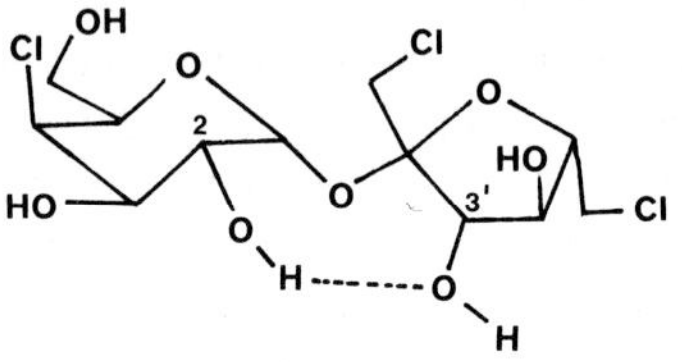

Neither 4-chloro-4-deoxy-D-galactose nor its glycoside derivatives were sweet (*152*) and attempts to enhance the sweetness of D-glucose, maltose and trehalose by replacing of selected hydroxyls by chlorine atoms have so far been unsuccessful (*153*). In fact 4,6:4′,6′-tetrachloro-4,6:4′,6′-tetradeoxy-*galacto*trehalose (**141**) was as bitter as quinine.

SIMPLE¹H-nmr spectroscopy (*154*) detected an intramolecular hydrogen bond in four 1′-chloro-1′-deoxy derivatives, including sucralose (**82**) with the 3′-hydroxyl as the donor and the 2-hydroxyl as the acceptor which gives a conformation (**142**) that is closely related to the **AH/B/X** glucophore. Hence the energy involved in initiating the sweet sensation by transforming this form to the intermolecular hydrogen bonded complex attached to the receptor protein will be minimal.

5. Natural Products Containing Sucrose

5.1. β-D-Fructofuranosyl Derivatives

Various β-D-fructofuranosyl derivatives of sucrose in which additional fructosyl units are attached at primary positions 1′,6′ and 6 to

References, pp. 175–184

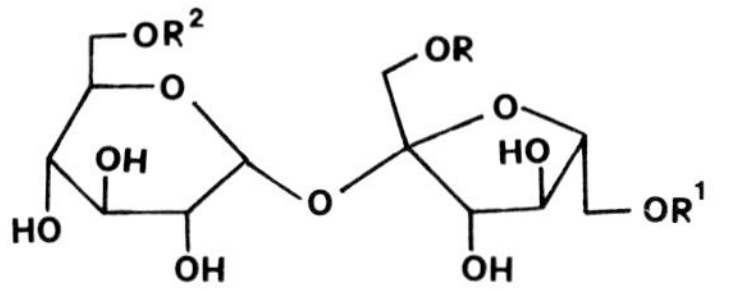

(143) R=β-D-Fruf, $R^1=R^2$=H
(144) R^1=β-D-Fruf, R=R^2=H
(145) R^2=β-D-Fruf, R=R^1=H

Fig. 37a. β-D-Fructopyranosyl and α-D-glucopyranosyl derivatives of sucrose

give trisaccharides and higher saccharides occur naturally. Many are of interest because of their reduced sweetness and lower nutritional value, whilst retaining the texture of sucrose in food products.

1-Kestose, 1'-O-(β-D-fructofuranosyl)sucrose, (143), has been isolated from some monocotylidons (155). Its synthesis in high yield has been achieved by using 50%, (w/v) sucrose solution and a fructosyl transferring enzyme from *A. niger* (156). The enzymic synthesis of a mixture of fructo-oligosaccharides containing 1-kestose as one of the main components was accomplished using an endotransferase from *Aureobacidium* sp (157).

6-Kestose, 6'-O-(β-D-fructofuranosyl)sucrose (144) is one of several fructo-oligosaccharides formed by the transferase action of yeast invertase on concentrated sucrose solutions (158, 159).

The same process yields, 6-O-(β-D-fructofuranosyl)sucrose (145) (160, 161). It has also been isolated from the sap of the sugar maple (162), and together with 1-kestose (143) from the aqueous alcoholic extracts of oat stalks (163). "Neosugar", a mixture of 1-kestose, the 1'-O-β-D-fructofuranosyl derivative of 1-kestose ("Nystose") and the 1'-O-β-D-fructofuranosyl derivative of Nystose, is being developed as a non-nutritive sweetener since it is not significantly metabolised by digestive enzymes in the gastro-intestinal tract and internal organs (164). These fructo-oligosaccharides occur in a variety of plants such as onion, asparagus root, tubers of Jerusalem artichoke and wheat, but Neosugar is manufactured from sucrose by use of a fungal fructosyltransferase. The product is 0.4–0.6 times as sweet as sucrose and has the advantage of characteristics similar to sugar in food products.

5.2. α-D-Glucopyranosyl Derivatives

Erlose, 4-O-(α-D-glucopyranosyl)sucrose (146), is formed by the action of honey invertase on sucrose (165), and is a promising low calorie and non-cariogenic sweetener (166).

6-O-(α-D-Glucopyranosyl) sucrose (**147**) is present in the roots of many species of gentian.

Melizitose (*167*), 3'-O-(α-D-glucopyranosyl)sucrose (**148**), is found in the "honey dew" of limes and poplars and as an exudate on Douglas

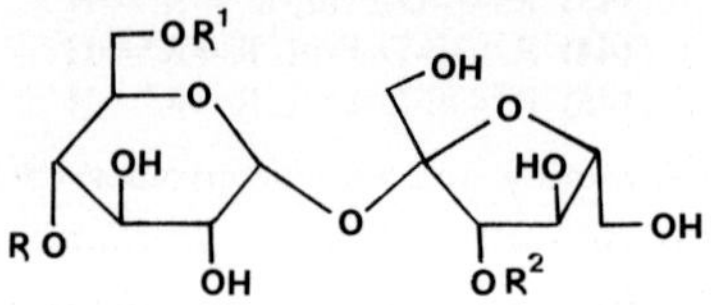

(**146**) R=α-D-Glcp, R^1=R^2=H
(**147**) R^1=α-D-Glcp, R=R^2=H
(**148**) R^2=α-D-Glcp, R=R^1=H

Fig. 37b

fir. The structure of melizitose has been ascertained by chemical (*168*) and biochemical (*169*) techniques and by X-ray diffraction studies (*170*). The crystal structure of melizitose monohydrate revealed that the two α-D-glucopyranosyl residues adopt the $^4C_1$ chair conformation whilst the fructofuranosyl unit assumed the $^4T_3$ conformation. The relative orientation of the D-glucose and D-fructose moieties was similar to that of sucrose.

5.3. α- and β-D-Galactopyranosyl Derivatives

Raffinose (*167*), 6-O-(α-D-galactopyranosyl)sucrose (**149**), is only slightly sweet (0.2 X) and is probably after sucrose the most abundant oligosaccharide of the plant world. The trisaccharide occurs in molasses of beet and cane sugar, and cotton seed meal (*171*). Cotton seed contains approximately 8% of raffinose and is not contaminated by monosaccharide and other oligosaccharides. The chemical synthesis of raffinose has been achieved by the condensation of 2,3,4,1',3',4',6'-hepta-O-acetylsucrose with tetra-O-benzyl-α-D-galactopyranosyl bromide in dry benzene in the presence of mercuric cyanide and Drierite followed by removal of protecting groups (*172*). Enzymically raffinose (**149**) is produced by the action of α-D-galactoside galactohydrolase, isolated from *Vicia faba* seeds, with melibiose and sucrose (*173*).

The crystal structure of raffinose pentahydrate revealed that the two pyranose rings are in the $^4C_1$ chair conformations and that the furanose ring adopts a puckered $^4T_3$ conformation (*174*). The orientation of the glycosidic linkage between the glucose and fructose units

References, pp. 175–184

is however different from that found in sucrose. Unlike sucrose, there are no intramolecular hydrogen bonds since all of the hydroxyl groups in raffinose act both as donors and acceptors of hydrogen bonds involving water molecules.

Isoraffinose, 6-*O*-(β-D-galactopyranosyl)sucrose (**150**) has been synthesised using a β-galactosidase from *E. coli* by the transgalactosylation from either *O*-nitrophenyl β-D-galactopyranoside or lactose in the presence of sucrose as an acceptor (*175*).

Lactosylfructoside, 4-*O*-(β-D-galactopyranosyl)sucrose (**151**), is formed by the action of the enzyme Levan-sucrase on a mixture of sucrose and lactose (*176*).

As its name suggests umbelliferose, 2-*O*-(α-D-galactopyranosyl)sucrose (**152**), is widely distributed amongst the Umbelliferae (*177*).

4'-*O*-(β-D-Galactopyranosyl)sucrose (**153**) has been isolated from the culture medium of *Leuconostoc mesenteroides* strain K (*178*).

Planteose, 6'-*O*-(α-D-galactopyranosyl)sucrose (**154**), is found in members of the Plantaginuceae, Labiateae, Solanaceae, and Mavaceae families (*179*) (*180*).

Stachyose, 6-*O*-(*O*-α-D-galactopyranosyl 1→6-α-D-galactopyranosyl)sucrose (**155**), is the second member of an homologous series of galactosylsucroses, raffinose (**149**) being the first member. This tetrasaccharide is found in many leguminous plants, in members of the Labiateae family, in conifers and in monocotyledons (*159*, *181*).

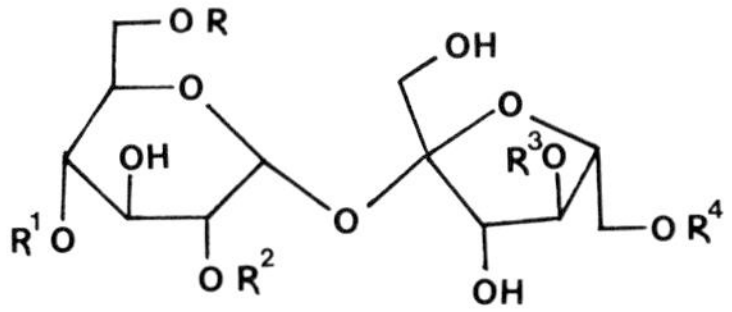

(**149**) R=α-D-Galp, $R^1=R^2=R^3=R^4=H$
(**150**) R=β-D-Galp, $R^1=R^2=R^3=R^4=H$
(**151**) R^1=β-D-Galp, $R=R^2=R^3=R^4=H$
(**152**) R^2=α-D-Galp, $R=R^1=R^3=R^4=H$
(**153**) R^3=β-D-Galp, $R=R^1=R^2=R^4=H$
(**154**) R^4=α-D-Galp, $R=R^1=R^2=R^3=H$
(**155**) R=α-D-Galp 1 → 6 α-D-Galp, $R^1=R^2=R^3=R^4=H$

Fig. 38. α- and β-D-Galactopyranosyl derivatives of sucrose

5.4. Galloyl Derivatives

Gallotannins contain a variety of carbohydrates such as glucose, fructose, alditols and glycosides. In 1988, an investigation of Chinese and North Korean rhubarbs revealed for the first time a range of galloyl monoesters of sucrose (*182*) (Fig. 39). On the basis of ^{1}H- and ^{13}C-nmr studies, FAB-MS combined with chemical and enzymic methods, their structures were assigned as 6-*O*-(**156**), 6'-*O*-(**157**), 4'-*O*-(**158**), 1'-*O*-(**159**), and 2-*O*-monogalloyl sucrose (**160**). Only the 4'-gallate (**158**) was not identified in the extracts of North Korean rhubarb.

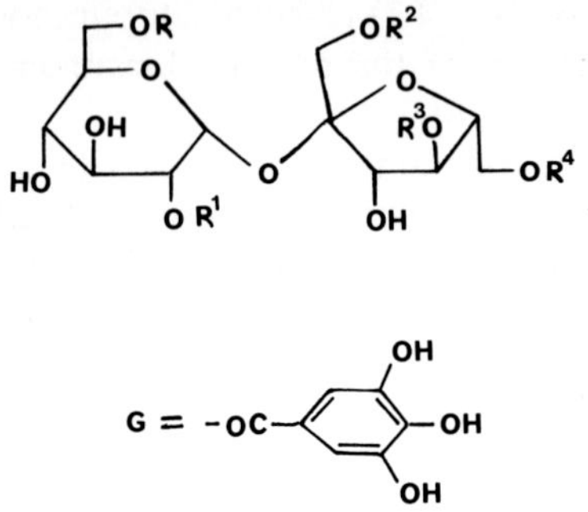

(**156**) R=G, R^1=R^2=R^3=R^4=H
(**160**) R^1=G, R=R^2=R^3=R^4=H
(**159**) R^2=G, R=R^1=R^3=R^4=H
(**158**) R^3=G, R=R^1=R^2=R^4=H
(**157**) R^4=G, R=R^1=R^2=R^3=H

Fig. 39. Galloyl esters of sucrose

5.5. Agrocinopine A

4'-*O*-Sucrose 2-*O*-L-arabinopyranosyl phosphate (**161**) has been isolated from dicotyledonous angiosperms infected with the gram-negative soil bacteria *Agrobacterium tumefaciens*. Its structure has been ascertained by ^{13}C, ^{1}H-, and 2D nmr studies (*183*) (*184*). A protected derivative of agrocinopine A (**162**) has been regiospecifically synthesised from 2,3,4,6,1',3',6'-hepta-*O*-acetylsucrose (**163**) and benzyl exo-3,4-*O*-benzylidene-β-L-arabinopyranoside (**164**) (*68*). Reaction of the sucrose ester (**163**) with 2,2,2-trichloroethylphosphorodichloridate in dry pyridine in the presence of 1,2,4-triazole gave the 4'-ester (**165**) which on treatment with the arabinoside (**164**) gave the desired phospho-diester derivative (**162**) (Fig. 40).

References, pp. 175–184

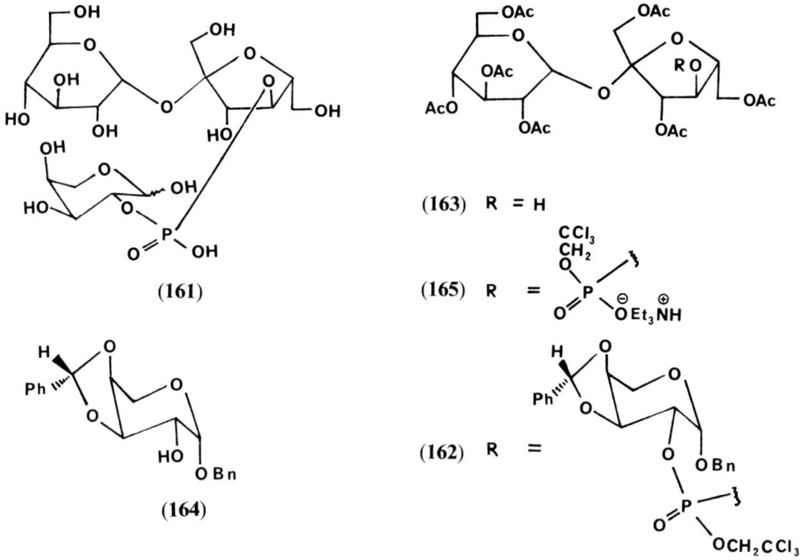

Fig. 40. Synthesis of Agrocinopine A

5.6. Sucrose Esters from Potato

The foliage of many wild potato species is covered with glandular trichomes (type A and B) that exude a chemical secretion which entraps anthropod pests. The exudates from type B trichomes of *Solanum berthaultii* Hawkes (P1 473340) and Hjerting (P1 498129) have been shown to contain sucrose esters as the major portion of the nonvolatile constituents (*185, 186, 187*). The esters were characterised by a combination of hydrolysis studies and ^{1}H-, ^{13}C- and 2D-nmr techniques as 6-*O*-capryl-3,4-di-*O*-isobutyryl-, 6-*O*-capryl-3,3′,4-tri-*O*-isobutyryl-, 6-*O*-capryl-3′-*O*-isobutyryl-3,4-di-*O*-(2-methylbutyryl)-, 2-*O*-acetyl-3′-*O*-hexanoyl-3,4-di-*O*-isobutyryl-, 2-*O*-acetyl-3′-4-di-*O*-hexanoyl-3-*O*-isobutyryl-, and 2-*O*-acetyl-3′-*O*-decanoyl-3,4-di-*O*-isobutyryl-sucrose.

5.7. Sucrose Esters from Tobacco

Sucrose esters are known to be flavour precursors in various tobaccos. One such ester, 6-*O*-acetyl-2,3,4-tri-*O*-(3*S*-methylpentanoyl)sucrose (**166**), has been isolated (*188*) and its structure confirmed by ^{1}H-, ^{13}C-nmr and mass spectrometry (*189*), and by unambiguous chemical

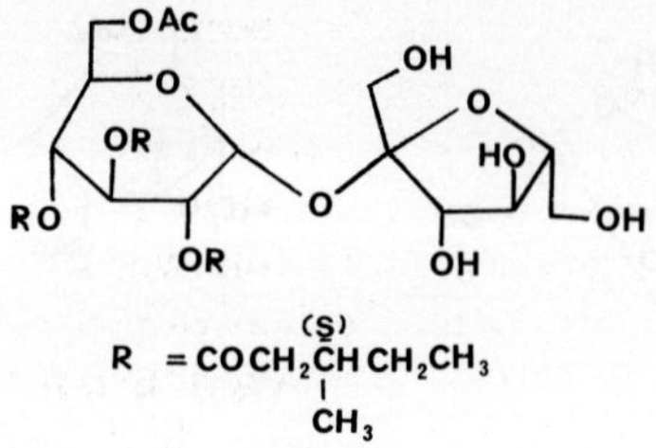

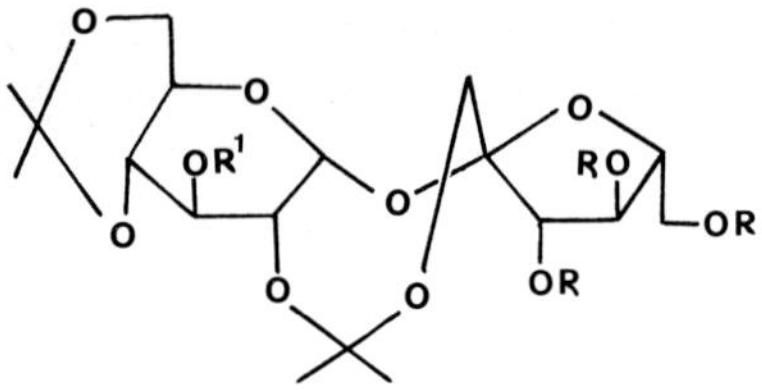

(34) R=R¹=Ac
(52) R=H, R²=Ac
a (167) R=Bn, R¹=CH₂CH=CH₂

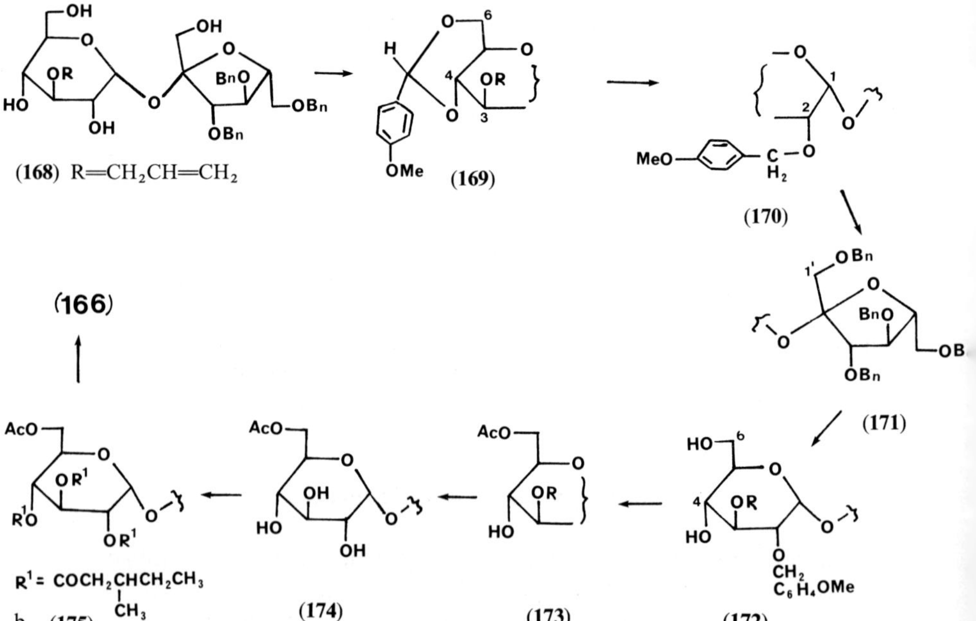

(168) R=CH₂CH=CH₂

(166)

R¹ = COCH₂CHCH₂CH₃
          |
          CH₃
b (175)

(174) (173) (172)

Fig. 41 a, b. Sucrose esters from Tobacco

ical synthesis (*190*). The ester has been elegantly synthesised from 4,6:1′,2-di-*O*-isopropylidene sucrose (Fig. 41). Thus selective deacetylation of the 3,3′,4′,6′-tetra-acetate (**34**) with sodium methoxide in methanol gave the 3-acetate (**52**) in 69% yield which on benzylation with benzyl bromide-silver oxide in DMF followed by conventional deacetylation and allylation afforded the 3-allyl-3′,4′,6′-tribenzyl ether (**167**) in high yield. Acid catalysed deacetylation furnished tetraol (**168**) which on treatment with 4-methoxybenzaldehyde dimethylacetal in DMF in the presence of toluene-*p*-sulphonic acid gave the 4,6-*O*-(4-methoxybenzylidene) acetal (**169**). Partial phase-transfer 4-methoxybenzylation of (**169**) then gave (**170**) with preferential substitution at the C-2 position, probably due to the 2-hydroxyl being more acidic than the 1′-hydroxyl group. The remaining 1′-hydroxyl group was then protected as a benzyl ether (**171**) to give the fully protected compound which was deacetalated to afford the 4,6-diol (**172**). This compound was converted into the 4,6-methoxyethylene derivative, using trimethylorthoacetate and toluene-*p*-sulphonic acid in acetonitrile, which was then converted under mild acidic conditions to the 6-acetate derivative (**173**). The 3-allyl and the 2-(4-methoxybenzyl) groups in (**173**) were removed by treatment with tris(triphenyl)rhodium (I) chloride and cerium (IV) ammonium nitrate, respectively, to give the 2,3,4-trihydroxy intermediate (**174**) which on acylation with (*S*)-3-methylpentanoic acid and thionyl chloride yielded ester (**175**), catalytic hydrogenolysis of which gave the target compound (**166**).

Acknowledgment

The authors wish express their sincere thanks and gratitude to Miss Pat Kenyon for considerable assistance in the production of the manuscript.

References

1. STRONG, L.A.G.: The Story of Sugar. London: George Weidenfold and Nicholson. 1954.
2. HUGILL, A.: Sugar and All That. London; Gentry Books, 1978.
3. LEVI, I., and PURVES, C.B.: Structure and Configuration of Sucrose. Adv. Carbohydr. Chem. **4**, 1 (1949).
4. HUDSON, C.S.: Relations between Rotatory Power and Structure in the Sugar Group. XXVI. The Ring Structures of Various Compound Sugars. J. Amer. Chem. Soc. **52**, 1707 (1930).
5. HAWORTH, W.N. and HIRST, E.L.: The Structure of Fructose, γ-fructose and Sucrose. J. Chem. Soc. 1858 (1926).
6. HAWORTH, W.N., HIRST, E.L., and LEARNER, A.: 1,3,4,6-Tetramethyl (γ-) Fructose and 2,3,5-trimethyl (γ-) Arabinose. Oxidation of d- and l-Trimethyl Arabinolactone. J. Chem. Soc. 2432 (1927).
7. FLEURY, P., and COURTOIS, J.E.: Oxidation of Sucrose by Periodic Acid. Compt. Rend. **214**, 366 (1942).

8. SCHLUBACH, H.H., and RAUCHALLES, G.: Die Spaltung des γ-Methylfructosids durch Saccharasen. Zur Konfiguration des Rohrzuckers. Ber. **58**, 1842 (1925).
9. BEEVERS, C.A., and COCHRAN, W.: The Crystal Structure of Sucrose Sodium Bromide Dihydrate. Proc. Roy. Soc. **A. 190**, 257 (1947).
10. PURVES, C.B., and HUDSON, C.S.: The Analysis of Gamma-methyl Fructoside Mixtures by Means of Invertase. IV. Behaviour of Sucrose in Methyl Alcohol containing Hydrogen Chloride. J. Amer. Chem. Soc. **56**, 1973 (1934).
11. JENNER, M.R.: In Developments in Food Carbohydrate-2. Ed., pp. 99–143. C.K. LEE. London: Applied Science Publishers Ltd. 1980.
12. KHAN, R.: Chemistry and New Uses of Sucrose: How Important? Pure and Appl. Chem. **56**, No. 7. 833–844 (1984).
13. Ref. 3. p. 27.
14. LEMIEUX, R.U., and HUBER, G.: A Chemical Synthesis of Sucrose. J. Amer. Chem. Soc. **75**, 4118 (1953).
15. TSUCHIDA, H., and KOMOTO, M.: A Modified Procedure for the Synthesis of Sucrose. Agr. Biol. Chem. **29**, 239 (1965).
16. NESS, R.K., and FLETCHER, H.G.: Synthesis of Sucrose and of α-D-Glucopyranosyl α-D-Fructofuranoside through the Use of 1,3,4,6-tetra-O-Benzyl-D-fructofuranose. Carbohydr. Res. **17**, 465 (1971).
17. QUEEN'S UNIVERSITY.: L-Sucrose. CANADIAN PATENT 1556 007 (1979).
18. ILEY, D., and FRASER-REID, B.: A New Synthesis of Sucrose which Demonstrates a novel Approach to the Synthesis of α-Linked Disaccharides. J. Amer. Chem. Soc. **97**, 2563 (1975).
19. DYER, Y.C., and KISHI, Y.: Synthesis of C-Sucrose. J. Org. Chem. **53**, 3383 (1988).
20. DEFAYE, J., DRIGUEZ, H., PONCET, S., CHAMBERT, R., and PETIT-GLATEON, M.F.: Synthesis of 1'-Thiosucrose and Anomers and the Behaviour of Levansucrase and Invertase with this Substrate Analog. Carbohydr. Res. **130**, 299 (1984).
21. LELOIR, L.F., and CARDINI, C.E.: The Biosynthesis of Sucrose Phosphate. J. Biol. Chem. **214**, 157 (1955).
22. NIKAIDO, M., and HASSID, W.Z.: Biosynthesis of Sacharides. Adv. Carbohydr. Chem. and Biochem. **26**, 366 (1971).
23. HASSID, W.Z., DOUDOROFF, M., and BARKER, H.A.: Enzymatically Synthesised Crystalline Sucrose. J. Amer. Chem. Soc. **66**, 1416 (1944).
24. HASSID, W.Z., DOUDOROFF, M., BARKER, H.A., and DORE, W.H.: Isolation and Structure of an Enzymatically Synthesised Crystalline Disaccharide, D-Glucosido-D-Ketoxyloside. J. Amer. Chem. Soc. **68**, 1465 (1946).
25. CARD, P.J., and HITZ, W.D.: Synthesis of 1'-Deoxy-1'-fluorosucrose via Sucrose Synthetase Mediated Coupling of 1'-Deoxy-1-fluorofructose with Uridine Diphosphate Glucose. J. Amer. Chem. Soc. **106**, 5348 (1984).
26. ZEMEK, J., and KUSAR, S.: Biosynthesis of Sucrose and its Deoxy Derivatives. Coll. Czech. Chem. Commun. **53**, 173 (1988).
27. BROWN, G.M., and LEVY, H.A.: Sucrose: Precise Determination of Crystal and Molecular Structure by Neutron Diffraction. Science **141**, 921 (1963); Further Refinement of the Structure of Sucrose Based on Neutron Diffraction Data. Acta crystallogr. Sect. B, **29**, 790 (1973).
28. BOCK, K., and LEMIEUX, R.U.: The Conformational Properties of Sucrose in Aqueous Solution: Intramolecular Hydrogen Bonding. Carbohydr. Res. **100**, 63 (1982).
29. MATHLOUTHI, M., LUU, C., MEFROY-BIGET, A.M., and LUU, D.V.: Laser-Raman Study of Solute-Solvent interactions in Aqueous Solutions of D-Fructose, D-Glucose and Sucrose. Carbohydr. Res. **81**, 213 (1980).
30. MCCAIN, D.C., and MARKLEY, J.L.: The Solution Conformation of Sucrose: Concentration and Temperature Dependence. Carbohydr. Res. **152**, 73 (1986).

31. CHRISTOFIDES, J.C., and DAVIES, D.B.: Comparison of Intramolecular Hydrogen-bonding Conformations of Sucrose Containing Oligosaccharides in Solution and the Solid State. Carbohydr. Res. **163**, 269 (1987).
32. MATHLOUTHI, M., SEUVRE, A.-M., and BIRCH, G.G.: Relationship between the Structure and the Properties of Carbohydrates in Aqueous Solutions; Sweetness of Chlorinated Sugars. Carbohydr. Res. **152**, 47 (1986).
33. SHAMIL S., BIRCH, G.G., and NJOROGE, S.: Intrinsic Viscosities and other Solution Properties of Sugars and their Possible Relation to Sweetness. Chemical Senses **13**, 457 (1988).
34. BIRCH, G.G., and SHAMIL, S.: Structure, Sweetness and Solution Properties of small Carbohydrate Molecules. J. Chem. Soc. Faraday Trans. 1, **84**, (8) 2635 (1988).
35. BELTON, P.S., and WRIGHT, K.M.: An ^{17}O-Nuclear Magnetic Resonance Relaxation-time Study of Sucrose-Water Interactions. J. Chem. Soc. Faraday Trans. *1*, **82**, 451 (1986).
36. HELFERICH, B.: Trityl Ethers of Carbohydrates. Adv. Carbohydr. Chem. **3**, 79 (1948).
37. JOSEPHSON, K.: Über Triphenylmethyl-äther einiger Di- und Trisaccharide. Ein Beitrag zur Kenntnis der Konstitution der Maltose, Saccharose und Raffinose. Ann. **472**, 230 (1929).
38. MCKEOWN, G.G., SERENIUS, R.S.E., and HAYWARD, L.D.: Selective Substitution in Sucrose. I. The Synthesis of 1',4,6'-Tri-*O*-methylsucrose. Can. J. Chem. **35**, 28 (1957).
39. MCKEOWN, G.G., and HAYWARD, L.D.: Selective Substitution in Sucrose. II. The Synthesis of 2,3,3',4,4'-Penta-*O*-methylsucrose and C_4 to C_6 Acetyl Migration in Sucrose. Can. J. Chem. **35**, 992 (1957).
40. OTAKE, T.: Studies of Tritylated Sucrose. I. Mono-*O*-tritylsucroses. Bull. Chem. Soc. Jpn. **45**, 3199 (1970).
41. OTAKE, T.: Studies on Tritylated Sucrose. II. Di-*O*-tritylsucroses. Bull. Chem. Soc. Jpn. **43**, 2895 (1972).
42. HOUGH, L., MUFTI, K.S., and KHAN, R.: Sucrochemistry. Part II. 6,6-Di-*O*-tritylsucrose. Carbohydr. Res. **21**, 144 (1972).
43. BUCHANAN, J.G., and CUMMERSON, D.A.: 1',4:3',6'-Dianhydrosucrose. Carbohydr. Res. **21**, 293 (1972).
44. BUCHANAN, J.G., CUMMERSON, D.A., and TURNER, D.M.: The Synthesis of Sucrose 6'-Phosphate. Carbohydr. Res. **21**, 283 (1972).
45. OTAKE, T.: Studies of Tritylated Sucroses. III. NMR Studies of Tritylacetylsucroses. Bull. Chem. Soc. Jpn. **47**, 1939 (1974).
46. KHAN, R., and MUFTI, K.S.: Synthesis and Reactions of 1',2:4,6-Di-*O*-isopropylidene Sucrose. Carbohydr. Res. **43**, 247 (1975).
47. BREDERECK, H., HAGELLOCH, G., and HAMBSCH, E.: Notiz zur Darstellung der Oktamethyl-saccharose. Chem. Ber. **87**, 35 (1954).
48. PERCIVAL, E.G.V.: Addition Compounds of the Carbohydrates. Part II. Potassium Hydroxide-Sucrose. J. Chem. Soc. **648** (1935).
49. LINDLEY, M.G., BIRCH, G.G., and KHAN, R.: Synthesis of Methyl Ether Derivatives of Sucrose. Carbohydr. Res. **43**, 360 (1975).
50. MOODY, W., RICHARDS, G.N., CHEETHAM, N.W.H. and SIRIMANNE, P.: Isolation and Alkaline Degradation of Some Mono-*O*-methyl Sucroses. Carbohydr. Res. **114**, 306 (1983).
51. MANLEY-HARRIS, M., and RICHARDS, G.N.: Studies of the Alkaline Degradation of Mono-*O*-methyl Sucroses. Carbohydr. Res. **90**, 27 (1981).
52. HENGLEIN, F.A., ABELSNES, G., HENEKA, H., LEINHARD, K., NAKMRE, P., and SCHEINOST, K.: Organosilyl Derivatives of Dicarboxylic Acids, Hydroxy acids and Sugars. Makromol. Chem. **24**, 1 (1957).

53. CHANG, C.D., and HASS, H.B.: Synthesis of a Silicone Derivative of Sucrose. J. Org. Chem. **23**, 773 (1958).
54. BENTLEY, R.: Preparation of Crystalline Octa-O-(trimethylsilyl)sucrose. Carbohydr. Res. **59**, 274 (1977).
55. FRANKE, F., and GUTHRIE, R.D.: t-Butyldimethylsilyl Ethers of Sucrose. Aust. J. Chem. **30**, 639 (1977).
56. FRANKE, F., and GUTHRIE, R.D.: 6,6'-Di-O-t-butyldimethylsilylsucrose: Studies on the Rearrangements Accompanying Deblocking of Such Silyl Ethers. Aust. J. Chem. **31**, 1285 (1978).
57. KARL, H., CEE, C.K., and KHAN, R.: Synthesis and Reactions of tert-Butyldiphenylsilyl Ethers of Sucrose. Carbohydr. Res. **101**, 31 (1982).
58. KHAN, R.: Sucrochemistry. Part XIII. Synthesis of 4,6-O-benzylidene Sucrose. Carbohydr. Res. **32**, 375 (1974).
59. KHAN, R., MUFTI, K.S., and JENNER, M.R.: Synthesis and Reactions of 4,5-Acetals of Sucrose. Carbohydr. Res. **65**, 109 (1978).
60. KHAN, R., and LINDSETH, H.: Selective Diacetalation of 1',2:4,6-di-O-isopropylidene Sucrose Tetra-acetate. Carbohydr. Res. **71**, 327 (1979).
61. CORTES-GARCIA, R., HOUGH, L., and RICHARDSON, A.C.: Acetalation of Sucrose by Acetal Exchange with Concomitant Fission of the Glycosidic Bond. J. Chem. Soc. Perkin I. 3176 (1981).
62. KHAN, R., JENNER, M.R., and JONES, H.F.: Synthesis and Reactions of Cyclic Acetal Derivatives of 6,6-dichloro-6,6'-dideoxysucrose. Carbohydr. Res. **49**, 259 (1976).
63. SCHÜTZENBERGER, P., and NAUDIN, M.: Sucre de canne et anhydride acétique. Bull. Chim. Soc. Fr. **12**, 206 (1869).
64. KONENKO, O.K., and KESTENBAUM, I.L.: Sucrose Monoacetate. J. Appl. Chem. **II**, 7 (1961).
65. KHAN, R., and MUFTI, K.S.: Process for the Preparation of 4,1',6'-trideoxy*galacto*-sucrose. U.K. patent 2079749 (1982).
66. BREDERECK, H., ZIMMER, H., WAGNER, A., FABER, G., GREINER, W., and HUBER, W.: Darstellung und Konstitution zweier Pentaacetyl-saccharosen. Chem. Ber. **91**, 2824 (1958).
67. BALLARD, J.M., HOUGH, L., and RICHARDSON, A.C.: Sucrochemistry. Part IV. A Direct Preparation of Sucrose 2,3,4,6,1',3',4'-Hepta-acetate. Carbohydr. Res. **24**, 152 (1972).
68. FRANZKOWIAK, L., and THIEM, J.: Synthesis of Agrocinopin A+B. Liebigs Ann. Chem. 1065 (1987).
69. CAPEK, K., VYDRA, T., RANNY, M., and SEDMERA, P.: Structures of Hexa-O-acetyl Sucroses Formed by Deacetylation of Sucrose Octaacetate. Coll. Czech. Chem. Commun. 2191 (1985).
70. CAPEK, K., VYDRA, T., and SEDMERA, P.: Structure of Penta-O-acetylsucroses Formed by Deacetylation of Octa-O-acetylsucrose. Reaction of 2,3,4,6,6'-Penta-O-acetylsucrose. Coll. Czech. Chem. Commun. **23**, 1317 (1988).
71. KHAN, R., JENNER, M.R., and LINDSETH, H.: Synthesis of Sucrose Epoxides; Partial De-esterification of 1',2:4,6-di-O-isopropylidene Sucrose Tetra-acetate and Selective Tosylation of 3,6'-di-O-Acetyl-1',2:4,6-di-O-isopropylidene Sucrose. Carbohydr. Res. **65**, 99 (1978).
72. AVELA, E., ASPELLIND, S., HOLMBOM, B., and MELANDER, B.: Selective Substitution of Hydroxyl Groups via Metal Chelates. In: Sucrochemistry, Ed. C.K. LEE, A.C.S. Symposium Series **41**, 62 (1977).
73. HOUGH, L., JAMES, C.E., KHAN, R., and RICHARDSON, A.C.: Unpublished Results.

74. CLODE, D.M., MCHALE, D., SHERIDAN, J.B., BIRCH, G.G., and RATHBONE, E.B.: Partial Benzoylation of Sucrose. Carbohydr. Res. **139**, 141 (1985).
75. CLODE, D.M., MCHALE, D., LAURIE, W.A., and SHERIDAN, J.B.: Partial Benzoylation of 2,1′:4,6-Di-O-Isopropylidene Sucrose. Carbohydr. Res. **139**, 147 (1985).
76. CLODE, D.M., LAURIE, W.A., MCHALE, D., and SHERIDAN, J.B.: Synthesis of 6,1′,3′-, 2,6,1′-, 1′,3′,6′- and 2,1′,6′-Tri-O-benzoyl Sucrose. Carbohydr. Res. **161**, 139 (1988).
77. OGAWA, T., and MATSUI, M.: A New Approach to the Regioselective Acylation of Polyhydroxy Compounds. Carbohydr. Res. **56**, C1 (1977).
78. BELORIZKY, R., EXCOFFIER, G., GAGNAIRE, D., UTILLE, J.P., VIGNON, M., and VOTTERO, P.: Synthese d'oligosaccharides sur polymere support. I – Le groupe β-benzoyl propionyle comme substituant temporaire. Bull. Soc. Chim. Fr. 4749 (1972).
79. GUTHRIE, R.D., LUCIAS, T.J., and KHAN, R.: The 4-Oxovaleryl and 3-Benzoylpropionyl Groups for the Protection of Hydroxyl Functions. Carbohydr. Res. **33**, 391 (1974).
80. HOUGH, L., CHOWDHARY, M.S., and RICHARDSON, A.C.: Selective Esterification of Sucrose using Pivaloyl Chloride. J. Chem. Soc. Chem. Commun. 664 (1978).
81. CHOWDHARY, M.S., HOUGH, L., and RICHARDSON, A.C.: Sucrochemistry. Part 33. The Selective Pivaloylation of Sucrose. J. Chem. Soc. Perkin I 419 (1984).
82. LEMIEUX, R.U., and BARRETTE, J.P.: A Chromatographic Analysis of the Product from the Tritosylation of Sucrose: Crystalline 6,6′-Di-O-tosylsucrose. Can. J. Chem. **38**, 656 (1960).
83. BOLTON, C.H., HOUGH, L., and KHAN, R.: New Derivatives of Sucrose Prepared from the 6,6′-di-O-tosyl- and the Octa-O-mesyl Derivatives. Carbohydr. Res. **21**, 133 (1972).
84. BRAGG, P.D., and JONES, J.K.N.: The Charaterisation of Tri-O-tosylsucrose. Can. J. Chem. **37**, 575 (1959).
85. BALL, D.H., BISSETT, F.H., and CHALK, R.C.: Synthesis of a 6,1′,6′-Tri-O-(mesitylenesulfonyl)-sucrose and a Further Examination of "Tri-O-tosyl-sucrose". Carbohydr. Res. **55**, 149 (1977).
86. BALLARD, J.M., HOUGH, L., PHADNIS, S.P., and RICHARDSON, A.C.: Selective Tetratosylation of Sucrose: Isolation of the 2,6,1′,6′-Tetrasulphonate. Carbohydr. Res. **83**, 138 (1980).
87. CREASEY, S.E., and GUTHRIE, R.D.: Mesitylenesulphonyl Chloride: a Selective Sulphonylating Reagent for Carbohydrates. J. Chem. Soc., Perkin I. 1373 (1974).
88. HOUGH, L., PHADNIS, S.P., and TARELLI, E.: The Direct Preparation of 1′,6,6′-Trideoxysucrose 1′,6,6′-Trimesitylenesulphonylsucrose. Carbohydr. Res. **44**, C12 (1975).
89. ALIQUIST, R.G., and REIST, E.J.: Synthesis of 6,6′-Disubstituted Sucrose Derivatives from 1-,6,6′-Tri-O-tripsyl sucrose. Carbohydr. Res. **46**, 33 (1976).
90. ISAACS, N.W., KENNARD, C.H., O'DONNELL, G.W., and RICHARDS, G.N.: X-ray Crystal Structure and Properties of a New Trianhydro-α-D-glucosyl 1,4:3,6-dianhydro-β-D-fructoside. J. Chem. Soc, Chem. Commun. 360 (1970).
91. HOUGH, L., and MUFTI, K.S.: Sucrochemistry. Part X. 1′,4,6′-Tri-O-mesylsucrose Pentaacetate: a Comparison of the Reactivity at the 4 and 1′ positions. Carbohydr. Res. **29**, 291 (1973).
92. CHUI, A.K.B., GURJAR, M.K., HOUGH, L., SINCHAROENKUL, L.V., and RICHARDSON, A.C.: The Synthesis of 2,1′-Anhydro-2,1′:3,6-Dianhydro and 2,1′:3,6:3′,6′-Trianhydro-sucrose. Carbohydr. Res. **100**, 247 (1982).
93. FAIRCLOUGH, P.H., HOUGH, L., and RICHARDSON, A.C.: Derivatives of β-D-Fructofuranosyl α-D-Galactopyranoside. Carbohydr. Res. **40**, 285 (1975).
94. GURJAR, M.K., HOUGH, L., RICHARDSON, A.C., and SINCHAROENKUL, L.V.: Preparation and Ring-opening of a 2,3-Anhydride derived from Sucrose. Carbohydr. Res. **150**, 53 (1986).

95. Hough, L., Chowdhary, M.S., and Richardson, A.C.: The use of Pivalic Esters for the Synthesis of Chloro, Azido and Anhydro Derivatives. Carbohydr. Res. **147**, 49 (1986).
96. Khan, R., Jenner, M.R., and Lindseth, M.: Synthesis of Sucrose Epoxides. Carbohydr. Res. **65**, 99 (1978).
97. Gurjar, M.R.: Further Transformations of Sucrose. Ph.D. Thesis. University of London (1980).
98. Hough, L., Kabir, A.K.M.S., and Richardson, A.C.: The Synthesis of Some Fluoro Derivatives of Sucrose. Carbohydr. Res. **125**, 247 (1984).
99. Helferich, B., Sprock, G., and Belser, E.: Über ein d-Glucose-5,6-dichlorohydrin. Ber. **58**, 886 (1925).
100. Buncel, E.: Chlorosulphates. Chem. Rev. **70**, 323 (1970).
101. Szarek, W.: Deoxyhalogeno Sugars. Adv. Carbohydr. Chem. **28**, 230 (1973).
102. Richardson, A.C.: Nucleophilic Replacement Reactions of Sulphonates. Part VI. A Summary of Steric and Polar Factors. Carbohydr. Res. **10**, 395 (1969).
103. Hough, L.: The Sweeter Side of Chemistry. Chem. Soc. Rev. **14**, 357 (1985).
104. Lee, C.K.: Synthesis of an Intensely Sweet Chlorodeoxysucrose: Mechanism of 4′-Chlorination of Sucrose by Sulphuryl Chloride. Carbohydr. Res. **162**, 53 (1987).
105. Parolis, H.: The Preparation of 4,6-Dichloro-4,6-dideoxy-α-D-galactopyranosyl 6-Chloro-6-deoxy-β-D-fructofuranoside 1′,2,3,3′,4′-Pentachlorosulphate. Carbohydr. Res. **48**, 132 (1976).
106. Ballard, J.M., Hough, L., Richardson, A.C., and Fairclough, P.H.: Sucrochemistry. Part XII. Reaction of Sucrose with Sulphuryl chloride. J. Chem. Soc. Perkin I 1524 (1973).
107. Khan, R.: Sucrochemistry. Part VII. Preparation and Reactions of Penta-O-benzoylsucrose 1′,6,6′-Tris(chlorosulphate) and Hexa-O-benzoylsucrose 6,6′-Bis(chlorosulphate). Carbohydr. Res. **25** 504 (1972).
108. Riva, S., Chopineau, J., Kieboom, A.P.G., and Klibanov, A.M.: Protease-catalyzed Regioselective Esterification of Sugars and Related Compounds in Anhydrous Dimethylformamide. J. Amer. Chem. Soc. **110**, 584 (1988).
109. Parkin, A., and Poller, R.C.: Sucrose Hydrogen Phthalates and Hydrogen Succinates. Carbohydr. Res. **62**, 83 (1978).
110. Theobald, R.S.: Carbonic Esters of Sucrose. Part II. The Polymerisation of O-Alkoxycarbonylsucroses. J. Chem. Soc. 5370 (1961).
111. Hough, L., Priddle, J.E., and Theobald, R.S.: The Carbonates and Thiocarbonates of Carbohydrates. Adv. Carbohydr. Chem. **15**, 91 (1960).
112. Kollonitsch, V.: "Sucrochemicals", Kline, The International Sugar Research Foundation, Washington D.C., p. 83 (1970).
113. Avenal, D., Neuman, A., and Gilliear-Pardraud, M.: X-ray Crystal Structure of Sucrose Octasulphate Heptahydrate. Acta Crystallogr. Sect. B **32**, 2598 (1976).
114. Khan, R.: Sucrochemistry. Part III. 1′,6,6′-Tri-O-tosylsucrose and its Conversion into 1′,4′:3,6:3′,6′-Trianhydrosucrose. Carbohydr. Res. **22**, 441 (1972).
115. Khan, R., Jenner, M.R., and Mufti, K.S.: Reaction of Methanesulphyl Chloride-N,N-Dimethylformaldehyde with Partially Esterified Derivatives of Sucrose. Carbohydr. Res. **39**, 253 (1975).
116. Lemieux, R.U., and Barrette, J.P.: 3,6-Anhydro-α-D-galactopyranosyl 1,4:3,6-Dianhydro-β-D-frunctoside-Configuration at the Anomeric Center of the Fructose Moiety of Sucrose. J. Amer. Chem. Soc. **80**, 2243 (1958).
117. Guthrie, R.D., Jenkins, I.D., Thang, S., and Yamasaki, R.: Derivatives of Sucrose 3′,4′-Epoxide. Carbohydr. Res. **121**, 109 (1983).
118. Capek, K., Vydra, T., and Sedera, P.: Oxirane-oxetane-1,4-dioxane Anhydro-ring Formation in Sucrose Derivatives. Carbohydr. Res. **186**, C1 (1987).

119. HOUGH, L., and MUFTI, K.S.: Sucrochemistry. Part VI. Further Reactions of 6,6'-Di-O-tosylsucrose and a Comparison of the Reactivity at the 6 and 6' positions. Carbohydr. Res. **25**, 497 (1972).
120. HOUGH, L., and MUFTI, K.S.: Mono-, Di-, Tri- and Tetrasubstituted Derivatives Prepared from Sucrose Octamethanesulphonate. Carbohydr. Res. **27**, 47 (1973).
121. GUTHRIE, R.D., and WATTERS, J.J: 1'-Derivatives of Sucrose and their Acid Hydrolysis. Aust. J. Chem. **33**, 2487 (1980).
122. HOUGH, L., PHADNIS, S.A., and TARELLI, E.: The Preparation of 4,6-Dichloro-4,6-dideoxy-α-D-galactopyranosyl 6-Chloro-6-deoxy-β-D-Fructofuranoside and the Conversion of Chlorinated Derivatives into Anhydrides. Carbohydr. Res. **44**, 37 (1975).
123. CASTRO, B., CHAPLEUR, Y., and GROSS, B.: Sels d'Alkyloxyphosphonium. Partie VII. Activation Selective de l'α,α-Trehalose et du Saccharose. Carbohydr. Res. **36**, 412 (1974).
124. KHAN, R., BHARDWAJ, C.L., MUFTI, K.S., and JENNER, M.R.: Synthesis of 6,6'-Dichloro-6,6'-dideoxysucrose Hexaacetate and Conversion into 6,6'-Diamino-6,6'-dideoxysucrose. Carbohydr. Res. **78**, 185 (1980).
125. ANISUZZAMAN, A.K.M., and WHISTLER, R.L.: Selective Replacement of Primary Hydroxyl Groups in Carbohydrates: Preparation of some Carbohydrate Derivatives containing Halomethyl Groups. Carbohydr. Res. **61**, 511 (1978).
126. HOUGH, L., PHADNIS, S.P., TARELLI, E., and PRICE, R.: The Application of ^{13}C-n.m.r. to Products Derived from Sucrose. Carbohydr. Res. **47**, 151 (1976).
127. HOUGH, L., KABIR, A.K.M.S., and RICHARDSON, A.C.: The Synthesis of 4-Deoxyfluoro and 4,6-Difluoro Derivatives of Sucrose. Carbohydr. Res. **131**, 335 (1984).
128. DRUCKER, D.B.: Comparative Effects of Five Chlorosucrose Analogues on Acidogenecity and Adherence of the Oral Bacterium *Streptococcus Mutans in vitro*. Arch. oral Biol. **28**, 833 (1983).
129. CHEN, C.C., WHISTLER, R.L., and DANIEL, J.R.: Synthesis of 6,6'-Dideoxysucrose. Carbohydr. Res. **117**, 318 (1983).
130. KHAN, R., and JENNER, M.R.: Synthesis and Reactions of Sucrose-5- and 5'-enes, Carbohydr. Res. **48**, 306 (1976).
131. TOUFEILI, I.A., DZIEDZIC, S.Z., and RATHBONE, E.B.: C-Methylation of Sucrose: Synthesis of 6- and 6'-C-Methylsucrose. Carbohydr. Res. **148**, 279 (1986).
132. CHUI, A.K.B., HOUGH, L., RICHARDSON, A.C., TOUFEILI, I.A., and DZIEDZIC, S.Z.: Synthesis of 4-C-Methyl and 4-C-Allyl Derivatives of Sucrose. Carbohydr. Res. **162**, 316 (1987).
133. KHAN, R., and PATEL, G.: Branched-chain Sucroses: Synthesis of 4,1',6'-Trichloro-4,1',4',6'-tetradeoxy-4'-C-methyl*galacto*sucrose. Carbohydr. Res. **162**, 298 (1987).
134. KHAN, R., and PATEL, G.: Branched-chain Sucroses: Synthesis and Wittig Reaction of the 1'-Aldehydro Derivative of Sucrose. Carbohydr. Res. **162**, 209 (1987).
135. KHAN, R., MUFTI, K.S., and PARKER, K.J.: Sucrose Derivatives. U.K. Patient 1 431 559 (1976).
136. UMEZAWA, S., TSUCHIYA, T., NAKADA, S., and TATSUTA, K.: Studies of Aminosugars. XIV. Syntheses of 6,6'-Diamino-6,6'-dideoxymaltosylamine, 1',6,6'-Triamino-1',6,6'-trideoxysucrose and 6,6'-Diamino-6,6'-dideoxytrehalose. Bull. Chem. Soc. Jpn. **40**, 395 (1967).
137. KHAN, R., MUFTI, K.S., and JENNER, M.R.: Sucrochemistry. Part XI. Synthesis of 1',6,6'-Triamino-1',6,6'-trideoxy Derivatives of Sucrose. Carbohydr. Res. **30**, 183 (1973).
138. SUAMI, T., IKEDA, T., NISHIYAMA, S., and ADACHI, R.: Sucrose Chemistry. 6. Synthesis of Amino Derivatives of Sucrose. Bull. Chem. Soc. Jpn. **48**, 1953 (1975).
139. KHAN, R., JENNER, M.R., LINDSETH, H., MUFTI, K.S., and PATEL, G.: Ring-opening

Rections of Sucrose Epoxides: Synthesis of 4'-Derivatives of Sucrose. Carbohydr. Res. **162**, 199 (1987).
140. MITRA, A.K., and PERLIN, A.S.: The Reaction of Sucrose with Glycol-cleaving Reagents. Can. J. Chem. **37**, 2047 (1959).
141. HALE, K.J., HOUGH, L., and RICHARDSON, A.C.: Morpholinoglucosides: New Potential Sweeteners Derived from Sucrose. Chem. and Ind. 268 (1988).
142. HALE, K.J., HOUGH, L., and RICHARDSON, A.C.: The Cyclisation of the Di- and Tetra-aldehydes Derived from Sucrose with Nitro-alkanes. Tetrahedron Letters **28**, 891 (1987).
143. HOUGH, L., SINCHAREONKUL, L.V., RICHARDSON, A.C., AKHTAR, F., and DREW, M.G.B.: Bridged Derivatives of Sucrose: the Synthesis of 6,6'-Dithiosucrose, 6,6'-Epidithiosucrose and 6,6'-Epithiosucrose. Carbohydr. Res. **174**, 145 (1988).
144. SHALLENBERGER, R.S., and ACREE, T.E.: Molecular Theory of Sweet Taste. Nature **480**, 216 (1967).
145. LEE, C.-K.: The Chemistry and the Biochemistry of the Sweetness of Sugars. Adv. Carbohydr. Chem. and Biochem. **45**, 199 (1987).
146. KIER, B.K.: A Molecular Theory of Sweet Taste. J. Pharm. Sci. **61**, 1394 (1972).
147. HOUGH, L., and KHAN, R.: Enhancement of the Sweetness of Sugar. In: Developments in Sweetness – 4. Ed., to be published T.H. Grenby. London-New York: Elsevier Applied Science Ltd. 1989.
148. HOUGH, L.: Sucrose, Sweetness and Sucralose. International Sugar J. **91**, 1062 (1989).
149. HOUGH, L., PHADNIS, S.P., KHAN, R., and JENNER, M.R.: Sweeteners. U.K. Patent 1 543 167 (1979).
150. HOUGH, L., and KHAN, R.: Intensification of Sweetness. T.I.B.S. 61 (1978).
151. KHAN, R., and JENNER, M.R.: Bittering Agents. U.K. Patent 2 057 561A (1980).
152. THELWALL, L.A.W. unpublished results.
153. DZIEDZIC, S.Z., and BIRCH G.G.: Structural Functions of Taste in the Sugar Series: Function of the γ-Atribute in the Sweet Glycophore. J. Sci. Food Agric, **32**, 283 (1981).
154. CHRISTOFIDES, J.C., DAVIES, D.B., MARTIN, J.A., and RATHBONE, E.B.: Intramolecular Hydrogen Bonding in 1'-Sucrose Derivatives Determined by SIMPLE ^{1}H NMR Spectroscopy. J. Amer. Chem. Soc. 5738 (1986).
155. BACON, J.S.D.: The Trisaccharide Fraction of some Monocotyledons. Biochem. J. **73**, 507 (1959).
156. HIKENDA, H., HIRAYAMA, M., and SUMI, N.: A Fructooligosaccharide-Producing Enzyme from *Aspergillus Niger* ATCC 20611. Agric. Biol. Chem. **52**, 1181 (1988).
157. SHINOHARA, S.: Process for Producing Fructo-oligosaccharase. European Patent. 188 047 (1986).
158. ALBON N., BELL, D.J., BLANCHARD, P.H., GROSS, D., and RUNDELL, J.T.: Kestose-A Trisaccharide Formed by Yeast Invertase. J. Chem. Soc. 24 (1953).
159. ASPINALL, G.O., PERCIVAL, E., REES, D.A., and RENNIE, M.: In Rodd's Chemistry of the Carbon Compounds, ed. S. COFFEY, Vol. 1F, p. 654. Amsterdam: Elsevier. 1967.
160. GROSS, D.G., BLANCHARD, P.H., and BELL, D.J.: A Trisaccharide formed from Sucrose by Yeast Invertase. J Chem. Soc. 1727 (1954).
161. BACON, J.S.D.: Oligofructosides. Bull Soc. Chim. Biol. 1441 (1960).
162. HAQ, S., and ADAMS, G.A.: Oligosaccharides from the Sap of Sugar Maple (*Ager Saccharum* Marsh). Can. J. Chem. **39**, 1165 (1961).
163. SCHLUBACH, H.H., and BERNDT, J.: Untersuchungen über Polyfructosen. LIX. Der Kohlenhydratstoffwechsel im Hafer. Ann. **647**, 41 (1961).
164. OKU, T., TOKUNAGA, T., and HOSOYA, N.: Non-digestibility of a New Sweetener, "Neosugar" in the Rat. J. Nutrition **114**, 1574 (1984).

165. WHITE, J.W., and MAHER, J.: α-Maltosyl β-D-Fructofuranoside: a Trisaccharide Enzymically synthesised from Sucrose. J. Amer. Chem. Soc. **75**, 1259 (1953).
166. TAKEUCHI, K., SATRAI, S., and MIYAKE, T.: Crystalline Erlose. U.K. Patent. 2168352A (1983).
167. RATHBONE, E.B.: Raffinose and Melezitose. In: Developments in Food Carbohydrate-2. Ed. C.K. LEE, p. 145. Applied Science Publishers Ltd. London (1980).
168. RICHTMYER, N.K., and HUDSON, C.S.: Melezitose Monohydrate and its Oxidation by Periodate. J. Org. Chem. **11**, 610 (1946).
169. HEHRE, E.J., and CARLSON, A.S.: Evidence on the Constitution of Melezitose Through Degradation to Sucrose by Bacterial Action. Arch. Biochem. Biophys. **36**, 158 (1952).
170. AVENEL, P., NEUMAN, A., and GILLIER-PANDRAUD, H.: Structure Crystalline du Melezitose Monohydrate. Acta Cryst. **B32**, 2598 (1976).
171. CLARK, E.P.: An Improved Method for Preparing Raffinose. J. Amer. Chem. Soc. **44**, 210 (1922).
172. SUAMI, T., OTAKE, T., NISHIMURA, T., and IKEDA, T.: Synthesis of Raffinose and an Isomer. Carbohydr. Res. **26**, 234 (1973).
173. PRIDHAM, J.B., and WALTER, M.W.: α-Galactosidase and Alkaline β-Fructofuranosidase Activity in *vicia faba*. Biochem. J. **92**, 20P (1964).
174. BERMAN, H.M.: The Crystal Structure of a Trisaccharide, Raffinose Pentahydrate. Acta Cryst. **B26**, 290 (1970).
175. SUYAMA, K., ADACHI, S., TAHA, T., SHOMA, T., HUANG, C.-J., and MOH, T.: Isoraffinose (6-β-galactosylsucrose) Synthesised by the Intermolecular Transgalactosylation Reaction *E. Coli* β-Galactosidase. Agric. Biol. Chem. **50**, 2069 (1986).
176. AVIGAD, G.: Enzymatic Synthesis and Characterisation of a New Trisaccharide, α-Lactosyl-β-fructofuranoside. J. Biol. Chem. 229, 121 (1957).
177. SVENDSON, A.B.: Die Verbreitung der Pflanzenfamilie der Umbelliferen. Acta Chem. Scand. **10**, 1500 (1956).
178. SUZUKI, M., and MEHRE, E.J.: Lactulosucrose (4-β-galactosylsucrose), a New Trisaccharide Synthesised by Cultures of *Leuconostoe Mesenteroides* strain K (NRRLB-1299). Arch. Biochem. Biophys. **105**, 339 (1964).
179. FRENCH, D., WILD, G.M., YOUNG, B., and JAMES, W.J.: Constitution of Planteose. J. Amer. Chem. Soc. **75**, 709 (1953).
180. FRENCH, D., YOUNGQUIST, R.W., and LEE, A.: Isolation and Crystallisation of Planteose from Mint Seeds. Arch. Biochem. Biophys. **85**, 471 (1959).
181. FRENCH, D., WILD, G.M., and JAMES, W.J.: Constitution of Stachyose. J. Amer. Chem. Soc. **75**, 3664 (1953).
182. KASHIWADA, Y., MONAK, G.-I., and NISHIOKA, I.: Galloylsucroses from Rhubarb. Phytochem. **27**, 1472 (1988).
183. RYDER, M.H., TATE, M.E., and JONES, G.P.: Agrocinopin A, a Tumour-Inducing Plasmid-Coded Enzyme Product; a Phosphodiester of Sucrose and L-Arabinose. J. Biol. Chem. **259**, 9704 (1984).
184. MESSEN, E., LENAERTS, A., MONTAGNE, M.V., BRUYN, A.D., JANS, A.W.H., and BINST, G.V.: ^{1}H and ^{31}P NMR Spectroscopy of Agrocinopine. J. Carbohydr. Chem. **5**, 683 (1986).
185. KING, R.R., PELLETIER, Y., SINGH, R.P., and CALHOUN, L.A.: 3,4-Di-*O*-isobutyryl-6-*O*-caprylsucrose: the Major Component of a Novel Sucrose Ester Complex. J. Chem. Soc. Chem. Commun. 1078 (1986).
186. KING, R.R., SINGH, R.P., and CALHOUN, L.A.: Isolation and Characterisation of 3,3′,4,6-Tetra-*O*-acylated Sucrose Esters from the Type B Glandular Trichomes of *Solanum Berthanetii*. Hawkes (P1 265957). Carbohydr. Res. **166**, 113 (1987).

187. KING, R.R., SINGH, R.P., and CALHOUN, L.A.: Elucidation of Structures for a Unique Class of 2,3,4,3'-Tetra-O-acylated Sucrose Esters from the Type B Glandular Trichomes of *Solanum Neocardenasii*. Hawkes and Hjerting (P1 498129). Carbohydr. Res. **173**, 235 (1988).
188. SEVERSON, R.F., ARRENDALE, R.F., CHORTYK, O.T., GREEN, C.R., THORNE, F.A., STEWART, J.L., and JOHNSON, A.W.: Isolation and Characterisation of the Sucrose Esters of the Cuticular Waxes of Green Tobacco Leaf. J. Agric. Food Chem. **33**, 870 (1985).
189. WAHLBERG, I., WALSH, E.B., FORSBLOM, I., OSCARSON, S., ENZEL, C.R., RYHAZE, R., and ISAKSSON, R.: Tobacco Chemistry 64. A New Sucrose Ester from Greek Tobacco. Acta Chem. Scand. **B40**, 724 (1986).
190. GAREGG, P.J., OSCARSON, S., and RITZEN, H.: Partially Esterified Sucrose Derivatives: Synthesis of 6-O-Acetyl-2,3,4-tri-O-[(s)-3-methylpentanoyl]sucrose, a Naturally Occuring Flavour Precursor of Tobacco. Carbohydr. Res. **181**, 89 (1988).

(Received April 14, 1989)

Author Index

Page numbers printed in *italics* refer to References

Abe, R. *86*
Abe, S. 5, *30*
Abelsnes, G. *177*
Achenbach, H. 11, *32, 33, 34*
Achmed, S. *32*
Ackermann, K. *81*
Acree, T.E. 165, *182*
Acton, E.M. *82*
Adachi, R. *181*
Adachi, S. *183*
Adams, G.A. *182*
Adams, P.E. 5, *30*
Adityachaudhury, N. 3, *29*
Ahmed, Z. *87*
Akai, S. *84, 86, 88*
Akhtar, F. *182*
Albon, N. *182*
Alemany, A. *32*
Alexander, J. *84*
Alfermann, A.W. *112*
Aliquist, R.G. *179*
Altintas, N. *85*
Ananthasubramanian, L. *79*
Anderson, D.K. *83*
Anderson, F.K. *82*
Angelucci, F. *85*
Anisuzzaman, A.K.M. *181*
Annoura, H. *84*
Anteunis, M.J.O. *31*
Arcamone, F. 38, *78, 82, 85*
Ardecky, R.J. *84, 85*
Arens, H. *109*
Argoudelis, A.D. 8, *31*
Arrendale, R.F. *184*
Ashcroft, A.E. *80*
Aspellind, S. *178*
Aspinall, G.O. *182*
Atta-Ur-Rahman *109*
Avela, E. 138, *178*

Avenal, D. *180*
Avenel, P. *183*
Avigad, G. *183*
Aweryn, B. *112, 114*
Ayer, W.A. *32*

Bacon, J.S.D. *182*
Baghdanov, V.M. *88*
Ball, D.H. *179*
Ballabio, M. *85*
Ballard, J.M. *178, 179, 180*
Balsevich, J. *110*
Baltus, W. *86, 87*
Balza, F. *33*
Barber, R.B. *86*
Barchielli, G. *85*
Barker, H.A. *176*
Barrette, J.P. *179, 180*
Bartels-Keith, J.R. 9, *32*
Bartmann, W. *108*
Barz, W. *109*
Basha, A. *109*
Bast, G. *111*
Basu, D. *33*
Bauman, J.G. *85, 86*
Baxter, R.L. *110, 114*
Bayer, E. *30*
Bayer, O. *78*
Beecham, A.F. 25, *33*
Beevers, C.A. *176*
Behnke, B. *78, 80*
Bell, D.J. *182*
Belleau, B. *85*
Belletire, J.L. *87*
Belorizky, R. *179*
Belser, E. *180*
Belton, P.S. *177*
Benfaremo, N. *87*
Bennani, F. *87*

Author Index

Bentley, R. *178*
Berlin, J. *110, 111, 112, 115*
Berman, H.M. *183*
Bernardi, L. *78*
Berndt, J. *182*
Bhardwaj, C.L. *181*
Biehl, E.R. *86*
Binst, G.V. *183*
Birch, A.J. 11, 12, *32*, 88
Birch, G.G. *177, 179, 182*
Bissett, F.H. *179*
Blackman, A.J. *32*
Blade, R.J. *83*
Blake, D.A. *108*
Blanch, H.W. *111*
Blanchard, P.H. *182*
Blomster, R.N. *110*
Blunden, G. *33*
Bock, E. *88*
Bock, K. 127, *176*
Boddy, I.K. *80*
Boeckman, R.K. Jr. *79, 80*
Boettger, S.D. *81*
Bohlmann, F. 5, 14, *30, 33*
Boily, Y. *31*
Bolton, C.H. *179*
Bonafede, J.D. *32*
Boniface, P.J. *80*
Borah, K. *79*
Borate, H.B. *81, 83*
Borremans, F. *31*
Botta, B. *114*
Boulet, C.A. *108, 114*
Bouyssou, H. *111*
Bragg, P.D. *179*
Brankiewicz, A.J. *29*
Brassard, P. *79*
Braun, M. 41, 57, *78*
Bray, B.L. *108*
Brazhnikova, M.G. *78*
Bredereck, H. *177, 178*
Brereton, R.G. 26, *34*
Breslow, R. 43, *79*
Brian, P.W. *29*
Broadhurst, M.J. 54, *78, 82*
Brockmann, H. 37, 38, *77*
Brodelius, P. *111, 113*
Broser, E. *80, 82*
Brown, G.M. *176*
Brown, J.R. *78*
Brussani, G.A. *85*
Bruyn, A.D. *183*

Buchanan, J.G. 130, *177*
Buchholz, M. *112*
Budzikiewicz, H. *34*
Buncel, E. *180*
Bunge, R.H. *29*
Buschi, C.A. *114*
Butler, D.N. 11, 12, *32*
Bycroft, B.W. *78*

Calderari, G. *82*
Calhoun, L.A. *183, 184*
Cambie, R.C. *80, 83*
Cammarato, L.V. *110*
Capek, K. 136, *178, 180*
Carbon, F.E. *82*
Card, P.J. 126, *176*
Cardellina, J.H. *34*
Cardini, C.E. *176*
Carew, D.P. *109, 112*
Carlson, A.S. *183*
Carney, R.E. *32*
Casazza, A.M. *78*
Castro, B. *181*
Cava, M.P. 71, *84, 85, 87, 88*
Cavill, G.W.K. *30*
Cee, C.K. *178*
Chagnon-Dubé, M. *31*
Chalk, R.C. *179*
Chambert, R. *176*
Chan, K.-S. *81*
Chandra, B.M. *81, 83*
Chang, C.D. *178*
Chapleur, Y. *181*
Chapple, C. *114*
Chatson, K.B. *109, 110, 111, 112, 114*
Chaykovsky, M. *82*
Cheetham, N.W.H. *177*
Chen, C.C. *181*
Chen, S. *82*
Chen, T.H.H. *111*
Chenard, B.L. *82*
Cheng, K.D. *112*
Chenieux, J.C. *111, 112*
Cherry, P.C. *110*
Chiong, K.G. *84*
Chmielewska, I. *29*
Choi, L.S.L. *108, 109, 110, 111, 112, 114*
Chopineau, J. *180*
Chortyk, O.T. *184*
Chowdhary, M.S. *179, 180*
Christofides, J.C. *177, 182*
Chui, A.K.B. *179, 181*

Author Index

Cieslak, J. 29
Clark, D.V. 30
Clark, E.P. *183*
Clarke, P.J. *34*
Clode, D.M. 138, *179*
Coburn, C.E. *83*
Cochran, W. *176*
Coffey, S. *182*
Coll, R.J. *79*
Collin, G.J. *83, 86*
Confalone, P.N. 40, *78*
Constabel, F. *109, 110, 111, 112, 113, 114*
Corcoran, R.J. *79*
Cordell, G.A. *108*
Corey, E.J. 52, 53, 60, *82*
Cortes-Garcia, R. *178*
Courtois, D. *109, 112*
Courtois, J.E. *175*
Crane, M.P. *83*
Craw, P.A. *80*
Creasey, S.E. *179*
Creasey, W.A. *108*
Crombie, L. 26, *34*
Crooke, S.T. *78*
Culos, K.O. *79*
Culver, M.G. *110*
Cummerson, D.A. 130, *177*
Curran, D.P. *83*
Curtis, P.J. *29*

Damratoski, D. *110*
Daniel, J.R. *181*
Das, A.K. 3, *29*
Dastur, K.P. *88*
Davies, D.B. *177, 182*
Davies, D.T. *80*
Davies-Coleman, M.T. 1, *31, 32*
Davis, J.B. *33*
Deb, T. *33*
De Capite, P. *110*
De Clercq, E. *87*
Deconti, R.C. *108*
Defaye, J. *176*
De Kimpe, N. *31*
Delgado, G. *32*
Del Nero, S. *82*
Deluca, V. *110, 113*
Deshpande, V.H. *83, 84, 86*
Desorby, V. *81*
Deus, B. *109, 110, 113*
Deus-Neumann, B. *110, 115*

Dhar, A.K. *33*
Dicosmo, F. *113*
DiMarco, A. *78*
Dimroth, O. *85*
Djerassi, C. *34*
Dodd, J.H. *88*
Doebner, O. *30*
Doireau, P. *112*
Dolak, T.M. *79*
Döller, G. *112*
Dolson, M.G. *82*
Dominguez, D. *84*
Domisse, R.A. *31*
Dore, W.H. *176*
Dorschel, C.A. *110*
Dötz, K.H. 49, *81*
Doudoroff, M. *176*
Dougan, J. *33*
Drapeau, D. *111*
Drew, M.G.B. *182*
Driguez, H. *176*
Drucker, D.B. *181*
Dubé, S. *31*
Ducos, J.P. *113*
Dyer, Y.C. *176*
Dziedzic, S.Z. *181, 182*

Ebeling, E. *87*
Echavarren, A. *79*
Eilert, U. *113*
El Khadem, H.S. *78*
Ellestad, G.A. 7, 8, *29, 31*
Ellis, G.P. *78*
El-Shagi, H. *109*
Elvidge, J.A. *34*
Endo, T. *114*
England, R.E. *33*
English, R.B. *32*
Enzel, C.R. *184*
Esmans, E.L. *31*
Evans, R.H. 8, *31*
Excoffier, G. *179*

Faber, G. *178*
Fahn, W. *110*
Fairclough, P.H. *179, 180*
Farina, F. *79, 86, 87*
Farnsworth, N.R. *108, 110*
Faust, T. *85*
Fayos, J. *32*
Fehlhaber, H.W. *30*
Felix, H. *113*

Author Index

Firth, P.A. 26, *34*
Fletcher, H.G. *176*
Fleury, P. *175*
Florent, J.-C. 50, *80, 82, 87*
Forsblom, I. *184*
Fowler, M.W. *111, 113*
Franca, N.C. 5, 30, *34*
Franke, F. *178*
Franzkowiak, L. 136, *178*
Fraser-Reid, B. *176*
French, D. *183*
French, J.C. 29, *32*
French, N.I. *84*
Frescos, J.N. *81, 83*
Fujimori, T. 30
Fujioka, H. *84*
Fujiwara, A.N. *84, 86*
Fukunaga, K. *86*
Funaishi, K. 33
Funayama, S. 33
Furukawa, S. *83*

Gagnaire, D. *179*
Galambos, J. 89
Gan, E.K. 33
Gandolfi, C. *82*
Gardner, J.M. 29, *33*
Gardner, J.N. 53, 57, *82*
Garegg, P.J. *184*
Garson, M.J. 26, *31, 34*
Gee, P.S. *82, 83*
Genot, A. 50, *82*
Gesson, J.-P. 44, *78, 79, 86, 87*
Giardino, P. *78*
Gigli, M. *85*
Gilbertson, S.R. *81*
Gillard, F. *32*
Gillard, J.W. *79*
Gillier-Pandraud, H. *180, 183*
Gioia, B. *85*
Girotra, N.N. *87*
Gleim, R.D. *78*
Gless, R.D. *86*
Glotter, E. *34*
Gnoj, D. *82*
Goerner, R.N. Jr. *79*
Golinski, J. *114*
Goodbody, A.E. *108, 114*
Gorczynska, K. 29
Gorgues, A. *85*
Gorter, K. 11, *32*
Goswami, S. 33

Gottlieb, O.R. *29, 34*
Govindachari, T.R. 18, *33*
Green, C.R. *184*
Greiner, W. *178*
Gröger, D. *109, 111*
Gross, B. *181*
Gross, D.G. *182*
Grove, J.F. 9, *32*
Grützmacher, H.-F. *87*
Gueritte, F. *109, 110*
Guern, J. *112*
Guillot, A. *112*
Gumulka, M. *114*
Gundlach, H. *110, 115*
Gupta, D.N. *87*
Gupta, R.C. *87*
Gurjar, M.K. *179, 180*
Gustowski, W. *114*
Guthrie, R.D. 152, 155, *178, 179, 180, 181*

Haag, A.P. *83*
Hagelloch, G. *177*
Hale, K.J. *182*
Hallmark, R.K. *84*
Hamann, I. *80*
Hambsch, E. *177*
Han, B.H. *108*
Hänsel, R. 33
Haq, S. *182*
Haque, W. *81, 88*
Harland, P.A. *87, 88*
Harrington, A.A. *78*
Harris, A.L. *109*
Harwood, L.M. *80*
Hasan, I. *79*
Hasan, M. *114*
Hashizume, T. 30
Hass, H.B. *178*
Hassall, C.H. 54, *78, 82*
Hassid, W.Z. *176*
Hata, N. 30
Hatakeyama, S. *79*
Hauser, F.M. 58, *80, 81, 83, 88*
Hauske, J.R. *32*
Hawley, R.C. *85*
Haworth, W.N. 119, *175*
Hayman, E.P. *112*
Haynes, L.J. *34*
Hayward, L.D. *177*
Heckenberg, U. *110*
Hegarty, P.K. *113*

Hehre, E.J. *183*
Heins, H. *80*
Heiskanen, J. *109*
Heissner, C.J. *31*
Helferich, B. 143, 146, *177, 180*
Hemme, C. *80*
Hemmi, S. *31*
Hemmi, T. *31*
Hemming, H.G. *29*
Heneka, H. *177*
Henglein, F.A. *177*
Henry, D.W. *84, 86*
Hermann, R. *85*
Hernandez, A. *32*
Herradon, B. *32, 35*
Herz, W. *31*
Hewitt, G.M. *114*
Hieke, M. *111*
Higashiyama, K. 63, *85*
Hikenda, H. *182*
Hirata, K. *108*
Hirata, T. *109, 110*
Hirayama, M. *182*
Hirst, E.L. 119, *175*
Hitz, W.D. 126, *176*
Hiyama, T. *84*
Hlubucek, J.R. 17, *33*
Ho, T. *82*
Hodge, P. *83, 87*
Hodgkinson, L.C. *80*
Hoffsommer, R.D. *87*
Höfle, G. *109, 110*
Hofmann, A.W. 4, *30*
Hokanson, G.C. *32*
Hokimoto, K. *29*
Hollenbeak K.H. *31*
Holmbom, B. *178*
Holmlund, C.E. *31*
Holum, J.R. *88*
Honda, T. *114*
Honek, J.F. *85*
Hoppe, H.J. *30*
Hori, K. *88*
Hosoya, N. *182*
Hough, L. 117, 136, 139, *177, 178, 179, 180, 181, 182*
Houk, K.N. *79*
Howe, T.A. *83*
Hoyer, T. *32*
Hsu, W.J. *112*
Huang, C.-J. *183*
Huang, Z.-D. *80*

Huber, G. 122, *176*
Huber, W. *178*
Hudson, C.S. 119, *175, 176, 183*
Hugill, A. *175*
Huhtikangas, A. *109, 110*
Hurley, T.R. *29*
Hutchinson, C.R. *110*
Huth, H. *34*

Ichimoto, I. 7, *31*
Iida, K. *30*
Iimori, T. *84*
Ikeda, T. *181, 183*
Iley, D. *176*
Imam, S.H. *78*
Inoue, S. *32*
Irvine, R.W. *79, 81, 82, 85, 88*
Isaacs, N.W. *179*
Isaksson, R. *184*
Ishizumi, K. *84*
Israili, Z.H. *34*

Jackson, D.A. *87*
Jackson, D.K. *82*
Jacquesy, J.C. *79, 86*
Jacquesy, J.-P. *87*
Jakupovic, J. *33*
James, C.E. 117, *178*
James, W.J. *183*
Jans, A.W.H. *183*
Janssen, N.J.M.L. *79*
Jarglis, P. *30*
Jarý, J. *30*
Javeed, S.M. *86*
Jenkins, I.D. *180*
Jenner, M.R. 122, *176, 178, 180, 181, 182*
Jerchel, D. 4, *30*
Jew, S. *82*
Jewers, K. *33*
Jiu, J. *31*
Johnson, A.W. *184*
Johnson, O.H. *88*
Jones, E.R.H. *34*
Jones, G.P. *183*
Jones, H.F. *178*
Jones, J.K.N. 144, *179*
Jones, P.G. *31*
Jones, P.S. *80*
Josephson, K. 129, *177*
Jung, M.E. *88*
Just, G. *32*

Kabir, A.K.M.S. *180, 181*
Kaiser, R. *30*
Kaji, K. *31*
Kakushima, M. *31*
Kaleya, R. *79*
Kallmerten, J. *42, 78, 79*
Kaneko, H. *30*
Karl, H. *178*
Karl, W. *34*
Kartha, K.K. *111*
Kashiwada, Y. *183*
Kasuga, R. *30*
Katagiri, K. *31*
Katsuno, T. *29*
Kauppinen, V. *109*
Kawamura, K. *33*
Kawasaki, M. *88*
Kazlauskas, R. *32*
Keay, B.A. *85*
Kelly, T.R. *78, 79, 86*
Kende, A.S. *57, 81, 83, 86, 88*
Kennard, C.H. *179*
Kerdesky, F.A.J. *85, 88*
Kestenbaum, I.L. *135, 178*
Khan, N. *87*
Khan, R. *117, 122, 130, 132, 135, 152, 176, 177, 178, 179, 180, 181, 182*
Khanapure, S.P. *86*
Khanna, I. *84*
Khanna, P.L. *79*
Kieboom, A.P.G. *180*
Kier, B.K. *166, 182*
Kikuchi, N. *30*
Kimura, Y. *7, 31, 78, 84, 85, 86*
King, P.F. *79*
King, R.M. *33*
King, R.R. *183, 184*
Kinoshita, M. *84*
Kirihara, M. *84*
Kirihata, M. *31*
Kirk, D.N. *34*
Kirson, I. *34*
Kishi, Y. *77, 79, 87, 176*
Kishimoto, H. *84, 88*
Kita, Y. *84, 86, 88*
Kjaer, A. *12, 30, 32*
Kleef, R.P. van *86*
Klesney, S.P. *30*
Klibanov, A.M. *180*
Klimars, M. *87*
Klingler, F.D. *30*
Knobloch, K.H. *111, 112*

Kobayashi, K. *33*
Koch, M. *87*
Koenig, F. *80*
Koga, K. *82*
Kohl, W. *109, 110*
Köhle, H.-J. *80, 87*
Kollonitsch, V. *180*
Kolodziejczyk, P. *109, 110, 111, 112, 114*
Komiyama, K. *33*
Komoto, M. *176*
Kondo, H. *84*
Konenko, O.K. *135, 178*
Kono, Y. *29, 33*
Kontnik, B. *29*
Korver, O. *32*
Koziski, K.A. *88*
Krauss, A.S. *79, 83, 85, 88*
Krauss, G. *111*
Kraychy, S. *31*
Krohn, K. *37, 78, 79, 80, 82, 85, 86, 87*
Krueger, R.J. *112*
Kuehne, M.E. *31*
Kuhn, R. *4, 30*
Kum, K. *30*
Kündig, E.P. *48, 81*
Kunstmann, M.P. *7, 8, 29, 31*
Kuo, C.H. *87*
Kurano, N. *108*
Kurten, U. *109*
Kurz, W.G.W. *109, 110, 111, 112, 113, 114*
Kusano, Y. *84*
Kusar, S. *176*
Kutney, J.P. *108, 109, 110, 111, 112, 114*
Kyi, A. *33*

Lakshmikantham, M.V. *85*
Lam, H.-Y. *81, 88*
Lamparsky, D. *30*
Langlois, N. *109*
Lapinjoki, S. *109, 110*
Larsen, D.S. *80, 83*
Laurie, W.A. *179*
Laussermair, E. *110*
Lavie, D. *34*
Lawrence, R.F. *108*
Learner, A. *175*
LeCoq, A. *85*
Lednicer, D. *84*
Lee, A. *183*

Lee, C.-K. *178, 180, 182*
Lee, G. *114*
Lee, K. *31*
Lee, S.L. *109, 110, 112*
Lee, W.W. *84, 86*
Leete, E. *34*
Lehne, V. *87*
Lehtola, T. *110*
Leinhard, K. *177*
Leloir, L.F. *176*
Le Men, J. *90, 108*
Lemieux, R.U. 122, 127, *176, 179, 180*
Lenaerts, A. *183*
Leung, N.L. *111*
Levi, I. *175*
Levy, H.A. *176*
Lewis, N.G. *114*
Li, T.-T. 58, *83*
Lichtenthaler, F.W. 4, *30, 35*
Lindley, M.G. *177*
Lindseth, H. *178, 180, 181*
Lombardi, P. *82*
Lorenz, K. *35*
Lounasmaa, M. 89, *108, 109, 110*
Lowe, J.A. *88*
Lown, J.W. *82, 88*
Lucias, T.J. *179*
Luckner, M. *109*
Lukěs, R. *30*
Luu, C. *176*
Luu, D.V. *176*
Lyding, J.M. *79*

Ma, W.Y. *35*
Mabry, T.J. *32*
MacCarthy, J.J. *112*
Mackenzie, N.E. *110*
Maeda, H. *86*
Maenosono, H. *83*
Magalhães, M.T. *29*
Maher, J. *183*
Majerus, F. *113*
Makmur, L. *32*
Mal, D. 58, *80, 83*
Manchanda, A.H. *33*
Mancini, M.L. *85*
Mandal, S.B. *82*
Manley-Harris, M. *177*
Marat, K. *81, 88*
Marigo, G. *111*
Markley, J.L. 127, *176*

Marks, M.E. *31*
Marquez, C. *32*
Marschalk, C. 45, *80*
Martin, J.A. *182*
Martinez, A.P. *86*
Martinez-Ripoll, M. *32*
Martin-Lomas, M. *32, 35*
Masaki, Y. 7, *31*
Mathlouthi, M. 127, *176, 177*
Matsjeh, S. *33*
Matsuda, F. *88*
Matsui, M. *179*
Matsumoto, T. *78, 86, 88*
Matsushita, H. *30*
Matter, U.E. *34*
Maurel, B. *113*
McCain, D.C. 127, *176*
McCallum, J.S. *81*
McDonald, H. *80*
McGahren, W.J. 7, *29, 31*
McHale, D. *179*
McHugh, M. *108, 114*
McInnes, A.G. *33*
McKeown, G.G. 129, *177*
McLauchlan, W.R. *114*
McNamara, J.M. *79, 87*
Meer, W.A. *110*
Mefroy-Biget, A.M. *176*
Mehendale, A.R. *81, 83*
Mehre, E.J. *183*
Meijr, T.M. 5, *30*
Meinwald, J. *34*
Melander, B. *178*
Merillon, J.M. *111, 112*
Messen, E. *183*
Meyer, H. 7, *31*
Meyer, H.H. 17, *33*
Meyers, A.I. 63, *85*
Mi, A.-Q. *81*
Miller, W.H. *86*
Mills, J.E. *83, 86*
Mincher, D.J. *80, 87*
Miqdad, M.T. *34*
Misawa, M. *108, 113, 114*
Mitra, A.K. *182*
Mitscher, L.A. *84*
Miura, Y. *108*
Miyakado, M. *32*
Miyake, T. *183*
Mizuba, S.S. *31*
Mizukami, H. *109, 110, 112*
Moh, T. *183*

Mohamed, A.L. *33*
Molina, T. *86*
Moll, G. *78*
Mollenschott, C. *115*
Monak, G.-I. *183*
Mondon, M. *78, 86*
Monneret, C. *50, 80, 82, 87*
Montagne, M.V. *183*
Moody, W. *177*
Moore, J.A. *83*
Moore, J.H. *31*
Moo-Young, M. *112*
Mori, K. *5, 30, 31*
Morley, J.O. *84*
Morris, P. *111, 112*
Morrow, G.W. *81, 83*
Mors, W.B. *3, 29*
Morton, G.O. *29*
Mosbach, K. *113*
Mothes, K. *109*
Mufti, K.S. 130, 135, *177, 178, 179, 180, 181*
Muir, A.D. *33, 34*
Mullen, P. *29*
Müller, G. *81*
Müller, U. *80, 87*
Murayama, T. *35*
Murphy, J. *82*
Murphy, P.T. *32*
Murphy, R.A. Jr. *87*
Murray, T.P. *31*

Naaranlahti, T. *109*
Naef, R. *82*
Naganawa, H. *84*
Nagata, K. *31*
Nakada, S. *181*
Nakahori, N. *33*
Nakajima, M. *82*
Nakano, J. *108, 114*
Nakata, T. *6, 30*
Nakmre, P. *177*
Narasimhan, L. *82*
Naudin, M. *178*
Nemec, J. *30*
Nemes, A. *108*
Ness, R.K. *176*
Neuman, A. *180, 183*
Neumann, D. *111*
Newmann, H. *88*
Nikaido, M. *176*
Nikaido, T. *114*

Nilsson, K. *111*
Nishimura, T. *183*
Nishioka, I. *183*
Nishiyama, S. *181*
Niskanen, M. *109*
Nivard, R.J.F. *79*
Njoroge, S. *177*
Nobuhara, A. 21, *34*
Noe, W. *115*
Noguchi, M. *30*
Nomura, K. *88*
Nordlöv, H. *110*
Norrestam, R. *30, 32*
Numata, A. *29*
Nylund, H. *109*

O'Connor, B. *32*
Oda, M. *31, 85*
O'Donnell, G.W. *179*
Ogawa, T. *179*
Ohashi, N. *84*
Ohe, H. *84*
Ohloff, G. *30*
Ohno, N. *32*
Ohsaki, M. *78, 88*
Ohta, K. *31*
Oishi, T. *30*
Okanishi, M. *33*
Okazaki, K. *88*
O'Krongly, D. *87*
Oku, T. *182*
O'Malley, G.J. *87*
Onodera, J. *114*
Oscarson, S. *184*
Otake, T. 129, 130, *177, 183*
Otsuka, T. *31*
Ouroussoff, N. *80*

Panchuk, B.D. *112*
Paredes, M.C. *86*
Pareilleux, A. *111, 113*
Parekh, N.D. *79*
Parker, K.A. 42, *78, 79, 88*
Parker, K.J. *181*
Parkin, A. *180*
Parolis, H. 146, *180*
Parthasarathy, P.C. 18, *33*
Pascual, C. *32, 34*
Patel, G. *181*
Patelli, B. *78*
Pauling, P.J. *34*
Paulser, M.G. *83*

Payne, G.F. *111*
Pearlman, B.A. *79*
Pelletier, Y. *183*
Pelter, A. *34*
Penco, S. 64, *85*
Perales, A. *32*
Percival, E. *182*
Percival, E.G.V. *177*
Pereda-Miranda, R. *32*
Perez, I. *114*
Perlin, A.S. *182*
Petiard, V. *109*
Petit-Glateon, M.F. *176*
Pfitzner, U. *115*
Phadnis, S.P. *179, 181, 182*
Pilley, B.A. *88*
Pirkle, W.H. 5, *30, 32*
Pitakowska, K. *29*
Pizzolato, G. *78*
Poling, S.M. *112*
Poller, R.C. *180*
Polonsky, J. 5, *30*
Poncet, S. *176*
Popall, M. *81*
Popein, D. *78, 82*
Potanova, N.P. *78*
Potier, P. *108*
Potman, F.J. *86*
Potman, R.P. *79*
Prados, P. *79, 87*
Prasanna, S. *80, 81, 88*
Pratesi, G. *78*
Preston, P.N. *84, 86*
Pretsch, E. *34*
Pribish, J.R. *86, 88*
Price, R. *181*
Priddle, J.E. *180*
Pridham, J.B. *183*
Priestap, H.A. 10, *32*
Priyono, W. *80, 87*
Procter, G. *35*
Pross, A. *34*
Puentes Saurez, A.M. *34*
Purves, C.B. *175, 176*
Puyvelde, L. van 6, *31*

Quesnel, A.A. *113*
Qureshi, A.A. *110*

Rabanal, R.M. *32, 35*
Radeloff, M. *80*
Rahm, M. *83*

Ralph, P.D. *34*
Ramakrishnan, G. *31*
Ranny, M. *178*
Rao, A.V.R. 50, *81, 83, 86*
Rao, B.R. *84, 86*
Rao, G.S.K. *83*
Rapoport, H. 63, *85, 86*
Ratcliffe, D. *112*
Rathbone, E.B. *179, 181, 182, 183*
Rauchalles, G. *176*
Raucher, S. *108*
Ravichandran, K. *84, 88*
Razak, D. *33*
Reddy, K.B. *83*
Reddy, K.R. *81*
Reddy, M.P. *83*
Reddy, N.L. *83*
Reddy, R.T. *86*
Rees, D.A. *182*
Regel, W. *33, 34*
Reich, S.D. *78*
Reinhard, E. *109, 112*
Reist, E.J. *179*
Ren, J. *81*
Rennie, M. *182*
Renoux, B. *79, 87*
Rhee, R.P. *88*
Richards, G.N. *177, 179*
Richardson, A.C. *178, 179, 180, 181, 182*
Richtmyer, N.K. *183*
Rideau, M. *111, 112*
Rieger, H. *82*
Riehl, J.J. *32*
Rigaudy, J. *30*
Ritzen, H. *184*
Riva, S. 148, *180*
Rivett, D.E.A. 1, *31, 32*
Rizzi, J.P. 57, *81*
Roberts, A.D. *78*
Robertson, A.V. 17, *33*
Robinson, C.W. *112*
Robinson, H. *33*
Rodrigo, R. *85*
Rodrigo, R.G.A. *108*
Roe, S.D. *113*
Romo de Vivar, A. *32*
Rosenbrook, W. *32*
Rosevear, A. *113*
Rozeboom, M.D. *79*
Rozynov, B.V. *78*
Rundell, J.T. *182*

Russell, A.T. *35*
Russell, R.A. *79, 81, 82, 83, 85, 86, 88*
Rutledge, P.S. *80, 83*
Ruveda, E.A. *32*
Ryder, M.H. *183*
Ryhaze, R. *184*

Sakata, I. *30*
Sakurai, T. *33*
Salisbury, P. *114*
Sam, T.W. *33*
Sarstedt, B. *87*
Sasaki, Y. *30*
Sasho, M. *84, 86, 88*
Sasse, F. *110, 112*
Sato, K. *5, 30*
Sato, T. *114*
Satrai, S. *183*
Savard, J. *79*
Sawahata, M. *84*
Saxton, J.E. *108*
Schallenberg, J. *111*
Schamp, N. *31*
Scheeren, H.W. *86*
Scheeren, J.W. *79*
Scheinker, Y.N. *78*
Scheinost, K. *177*
Schiel, O. *111*
Schiess, P. *82*
Schlubach, H.H. *176, 182*
Schmalle, H.W. *87*
Schoor, O. van *31*
Schuster, A. *33*
Schutte, H.R. *109*
Schützenberger, P. *178*
Schwenk, R. *78, 82*
Scieckus, V. *85*
Scott, A.I. *109, 110, 112, 114*
Scragg, A.H. *113*
Sedmera, P. *178, 180*
Seebach, D. *7, 31, 51, 82*
Sekihachi, J. *88*
Sekizaki, H. *79*
Semmelhack, A. *49, 81*
Sercel, A.D. *81, 82*
Serenius, R.S.E. *177*
Serizawa, Y. *31*
Seuvre, A.-M. *177*
Severson, R.F. *184*
Sew-Yeu, C. *33*
Shallenberger, R.S. *165, 182*
Shamil, S. *177*

Shaw, G. *68, 80, 87*
Sheldrick, W.S. *109*
Sheridan, J.B. *179*
Shibassaki, M. *84*
Shinohara, S. *182*
Shoma, T. *183*
Shuler, M.L. *111*
Sibi, M.P. *85*
Siegel, S.M. *29*
Sih, C.J. *78*
Simmons, D.P. *81*
Simon, W. *34*
Sincharoenkul, L.V. *179, 182*
Singh, M. *114*
Singh, R.P. *183, 184*
Sirimanne, P. *177*
Slates, H.L. *87*
Sleigh, S.K. *109, 110, 111, 112*
Smart, N.J. *113*
Smissman, E.E. *34*
Smith, J.I. *113*
Smith, L. *111*
Smith, S. *34*
Smith, T.H. *84, 86*
Smitka, T.A. *29*
Snatzke, G. *23, 25, 30, 33, 34*
Snider, B.B. *79*
Snieckus, V. *87*
Sodeoka, M. *84*
Soll, H.J. *109*
Sondhi, S.M. *82, 88*
Soranzo, C. *78*
Spencer, G.F. *33*
Spero, D.M. *88*
Sprock, G. *180*
Stafford, A. *111*
Stampwala, S.S. *29*
Stark, W. *21, 34*
Staunton, J. *26, 31, 34*
Steinman, C.E. *29*
Stephan, D. *85*
Sterns, M. *86*
Stewart, J.L. *184*
Stillwell, M.A. *31*
Stöckigt, J. *109, 110, 115*
Stoffregen, A. *87*
Stoodley, R.J. *66, 78, 87*
Strong, L.A.G. *175*
Strubin, T. *82*
Strunz, G.M. *7, 31*
Stuart, K.L. *109, 110, 111, 112, 114*
Suami, T. *181, 183*

Suarato, A. 85
Suen, R. 114
Sugiura, Y. 33
Sugiyama, T. 35
Suide, H. 31
Sum, F.-W. 80
Sumi, N. 182
Sutherland, J.K. 71, 80
Suwita, A. 5, 30
Suyama, K. 183
Suzuki, F. 78
Suzuki, M. 78, 84, 85, 86, 183
Suzuki, Y. 29, 33
Svendson, A.B. 183
Svoboda, G.H. 108
Swenton, J.S. 47, 81, 82, 83
Szarek, W. 180

Taha, T. 183
Takahashi, Y. 84, 85
Takemura, T. 29
Takeuchi, K. 183
Takeuchi, S. 29, 33
Takeuchi, T. 84
Talapatra, B. 33
Talapatra, S.K. 33
Tallevi, S.G. 113
Tamirez, J. 43, 79, 87
Tamoto, K. 85
Tamura, Y. 84, 86, 88
Tamuro, S. 31
Tanabe, Y. 32
Tang, P.-C. 81
Tanno, N. 79, 84
Tarelli, E. 179, 181
Tate, M.E. 183
Tatsuta, K. 84, 181
Tatum, J.H. 29
Taub, D. 87
Taylor, W.I. 90, 108
Tegmo-Larsson, L.-M. 79
Te Raa, J. 78
Terashima, S. 54, 78, 79, 82, 84, 85, 86, 88
Terashima, T. 88
Thang, S. 180
Thelwall, L.A.W. 182
Theobald, N. 34
Theobald, R.S. 180
Thiem, J. 136, 178
Thomas, G.J. 54, 78, 82
Thomas, S.A. 34

Thomson, R.H. 77
Thorne, F.A. 184
Toivonen, L. 109
Tokunaga, T. 182
Tolkiehn, K. 79, 87
Tomioka, K. 82
Toufeili, I.A. 181
Towers, G.H.N. 33, 34
Trachtenberg, E.N. 79
Tracy, M. 82
Treimer, J.F. 109, 110
Trenbeath, S. 78
Trost, B.M. 67, 87
Tsang, W.-G. 79, 86
Tsay, Y. 83, 86
Tschesche, R. 4, 30
Tse, A. 33
Tsubaki, K. 31
Tsuchida, H. 176
Tsuchiya, T. 181
Tsukamoto, H. 108, 114
Turner, D.M. 177
Tyler, R.T. 110, 112

Ueda, H. 31
Ughetto-Monfrin, J. 80
Umezawa, H. 84
Umezawa, I. 33
Umezawa, S. 181
Unwin, C.H. 29
Urbach, G. 21, 34
Utille, J.P. 179
Uwera, C. 31
Uwimana, E. 31

Valverde, S. 13, 32, 35
Vanotti, E. 85
Vaya, J. 79
Vedejs, E.V. 86, 88
Veith, R. 78
Venkatswamy, G. 86
Veräjänkorva, H. 109
Veysoglu, T. 84
Vigevani, A. 85
Vignon, M. 179
Villar, J.D.F. 32
Vinas, R. 113
Vlasova, T.F. 78
Vlietinck, A.J. 31
Vogel, P. 43, 66, 79, 87
Vottero, P. 179
Vukovic, J. 108, 114

Vydra, T. *178, 180*

Wada, A. *86*
Wagner, A. *178*
Wahlberg, I. *184*
Walsgrove, T.C. *83*
Walsh, E.B. *184*
Walter, M.W. *183*
Wan, W. *110*
Warrener, R.N. *81, 82, 83, 85, 86, 88*
Watabe, T. *85*
Watanabe, K. *32*
Watanabe, M. *83*
Watters, J.J. 155, *181*
Weber, S. *87*
Weiler, E. *109*
Weinreb, S.M. *88*
Wells, R.J. *32*
Welzel, P. *79*
Wendler, N.L. *87*
West, G.B. *78*
Wetchapinan, S. *33*
Whistler, R.L. *181*
White, J.W. *183*
Whitfield, F.B. *30*
Wielogorski, Z. *84*
Wild, G.M. *183*
Wilke, C.R. *111*
Williams, D.H. *34*
Williams, D.J. *87*
Willmer, N.E. *29*
Winterfeldt, E. *108*
Winwick, T. *84, 86*
Wiseman, J.R. *84*
Witte, B. *109, 110*
Wittmann, G. *34*
Witzke, J. 11, *32, 33*

Wolf, R.B. *33*
Wolgemuth, R.L. *84*
Wong, C.-M. 39, 49, 52, *78, 81, 82, 88*
Woo, W.S. *108*
Woodgate, P.D. *80, 83*
Worth, B.R. *109, 110, 111, 112, 114*
Wright, J.M. 29
Wright, K.M. *177*
Wu, H.Y. *84*
Wu, Y.L. *83*
Wulff, G. *30*
Wulff, W.D. 49, *81*

Yadav, J.S. *83*
Yamamoto, H. *84*
Yamamoto, I. *31*
Yamamoto, K. 29
Yamamoto, S. *31*
Yamano, T. *31*
Yamasaki, R. *180*
Yamashita, K. 35
Yang, D.C. *81*
Yang, G.-S. *81*
Yoahii, E. *88*
Yokoyama, H. *112*
Yoshida, E. *33*
Yoshida, S. *33*
Yoshioka, H. *32*
Young, B. *183*
Youngquist, R.W. *183*

Zbaskiy, V.B. *78*
Zemek, J. *176*
Zenk, M.H. *109, 110, 113, 115*
Zieserl, J.F. 8, *31*
Zimmer, H. *178*

Subject Index

Absidia glauca 28
Acanthoaustralide 14, 15
Acanthospermum australe 14
Acetaldehyde 8
3-Acetamido-3-deoxy-α-D-glucopyranosyl 4-acetamido-4-deoxy-β-D-glucoheptulopyranoside 163
Acetic acid 74, 129, 135, 139, 142, 152
Acetic anhydride 75, 135, 138, 160
Acetobromoglucose 67
Acetonitrile 157, 175
2-Acetoxybutadiene 65
19-Acetoxy-11-hydroxytabersonine 91, 94
19-Acetoxy-11-methoxytabersonine 91, 94
Acetylacetone 49
2-O-Acetyl-3′-O-decanoyl-3,4-di-O-isobutyryl-sucrose 173
2-O-Acetyl-3′-4-di-O-hexanoyl-3-O-isobutyryl-sucrose 173
2-O-Acetyl-3′-O-hexanoyl-3,4-di-O-isobutyryl-sucrose 173
2-Acetylnaphthalene 56
3′-O-Acetylsucrose 166
6-O-Acetylsucrose 135
Acetyltetralins 49
3-S-Acetyl-3-thio-altropyranoside 164
6-O-Acetyl-2,3,4-tri-O-(3S-methylpentanoyl)sucrose 173
Acidogenesis 157
Acute leukemia 38
Aglycones 77
Agrobacterium tumefaciens 172
Agrocinopine A 136, 172, 173
Agrocinopine B 136
Ajmalicine 90, 91, 92, 96, 98, 104
Aklavinones 43, 44, 47, 63, 64, 71
Akuammicine 91, 95
Akuammigine 91, 92
Akuammiline 91, 93
Alditols 171
Alkaline hydrolysis 9

6-Alkenyl-5,6-dihydro-α-pyrones 10, 11, 12
Alkoxybutadienes 44
α-Alkoxy esters 51
6-Alkyl-5,6-dihydro-α-pyrones 4
6-Alkyl-α-pyrones 3
Allyl bromide 49, 59, 69
Allyl magnesium bromide 158
Aloe-emodin 67
Alternaria citri 2, 15
Alternaria solani 9
Alternaric acid 9, 10, 28
Aluminum amalgam 52
Aluminum oxide 154
Amaryllidaceae 10
α-Aminobutyric acid 46
(S)-2-Aminobutyric acid 69
9-Aminodaunomycinones 60
4-Amino-4-deoxygalactosucrose 160
6′-Amino-6′-deoxysucrose 160
1-Amino-2,3-dihydroxy-propane 161
Ammonium chloride 160, 164
Ammonium thiocyanate 164
Anamarine 12, 13, 24
2,1′-Anhydro-1′-methoxycarbonylmethyl-sucrose 158
3,6-Anhydrosucrose 151
3′,4′-Anhydrovinblastine 107
Aniba gigantigolia 19
Anionic reactions 45
Annonaceae 4, 17
Anthracyclines 37, 38, 42, 43, 65
Anthracyclinones 37, 39, 45, 46, 49, 57, 63, 65, 69, 72, 77
Anthradiquinone 65
Anthrahydroquinones 67, 68
Anthraquinone derivatives 65
Anthraquinones 37, 39, 45, 46
1,4-Anthraquinones 65
9,10-Anthraquinones 70, 76
Anti-acidogenic activity 157

Antibacterial activity 28
Antibiotic U-13,933 8
Antifungal activity 2, 9, 28
Anti-inflammatory activity 29
Antileukemic drugs 90
Antimycotic activity 29
Antioedemic activity 29
Antipyretic activity 29
Antirhine 91, 95
Antitumor activity 2
Antitumor agents 14
Antitumor combination chemotherapy 38
Antitumor drugs 38
Aqueous acetic acid 129
L-Arabinose 123
Argentilactone 10, 11, 20, 24
Aristolochia argentina 10
Aristolochiaceae 10
Aroma components 4
6-Aryl-5,6-dihydro-α-pyrones 3, 16, 17, 18, 23
6-Aryl-α-pyrones 3
Asparagus root 169
Aspartame 166
Aspergillus caespitosus 8
Aspergillus carneus 8
Aspergillus elegans 9
Aspergillus melleus 9, 26
Aspergillus nidulans 8
Aspergillus niger 169
Aspergillus ochraceus 9
Aspergillus sp. 9
Aspergillus sp. NRRL 5769 8
Asperline 8, 9, 24, 25
Aspyrone 9, 24, 25, 26, 27
– Biosynthesis 27
Auramycinone 71
Aureobacidium sp. 169
Australian ants 5
3-Azido-3-deoxy-α-D-altropyranosyl β-D-fructopyranoside 161
4′-Azido-4′-deoxysucrose 161

Bacillus subtilis 125
Baker's yeast 62
Benzaldehyde dimethyl acetal 133
Benzocyclobutenediones 52, 73
Benzoic acid 17
Benzophenone 40
Benzoquinone 58, 59
Benzoyl chloride 138, 139
Benzoylpropionates 139

3-Benzoylpropionic acid 139
Benzyladenine 102
Benzylamine 161
Benzyl exo-3,4-O-benzylidene-β-L-arabinopyranoside 172
4,6-O-Benzylidenesucrose 143
4,6-O-Benzylidene sucrose hexa-acetate 124, 132, 142
2-O-Benzyl-3,4-O-isopropylidene-D-erythrose 124, 125
Berberidaceae 10
Biological activity 1, 2, 28, 159
Biosynthesis 2, 26, 27
– Aspyrone 27
– Psilotin 27
– 6-Substituted 5,6-dihydro-α-pyrones 26
Birch reduction 56, 75
Bis-thiouronium salt 163
Bis(tributylstannyl)-oxide 139
1,3-Bis(trimethylsiloxy)-1,3-butadiene 66
Bistrimethylsiloxy diene 43, 44
Boric acid 40
Boronolide 5, 6, 20, 24, 25
Boron tribromide 55
Boron trifluoride etherate 55
Boron trifluoroetherate 50
Brigl's anhydride 122, 123
N-Bromoacetamide 63
Bromoacetic ester 49
6′-Bromo-6′-deoxy-4-O-mesylsucrose 156
Burley tobacco 5
Butanone 147, 154
t-Butyldimethylsilyl chloride 131
6′-O-t-Butyldiphenylsilyl sucrose 131, 139
sec-Butyl lithium 74

Calcium alginate 106
Camphor sulphonic acid 124
6-O-Capryl-3,4-di-O-isobutyryl-sucrose 173
6-O-Capryl-3′-O-isobutyryl-3,4-di-O-(2-methylbutyryl)-sucrose 173
6-O-Capryl-3,3′,4-tri-O-isobutyryl-sucrose 173
Carbon source 101
Carbon dioxide 105
Carbon tetrabromide 156
Carbon tetrachloride 156
Carminomycin 38
Catharanthine 91, 94, 104, 106, 107

Subject Index

Catharanthus roseus 89, 90, 91, 95, 97, 98, 99, 100, 102, 103, 104, 105, 106, 107, 108
CD data 24
– 6-Substituted 5,6-dihydro-α-pyrones 24
Ceric ammonium nitrate 52, 61
Cerium (IV) ammonium nitrate 175
m-Chlorobenzoic acid 124
6-Chloro-6-deoxy-4-*O*-formylsucrose 156
4-Chloro-4-deoxy-D-galactose 168
4-Chloro-4-deoxy-galactosucrose 156, 157, 166
1'-Chloro-1'-deoxysucrose 155
6-Chloro-6-deoxysucrose 156, 166
6-Chloro-6-deoxysucrose hepta-acetate 151
6-Chloro-6-deoxy-6'-*O*-tosylsucrose hexa-acetate 154
Chloroform 135, 136, 144, 147
m-Chloroperbenzoic acid 8
Chloroprene 59, 65
Chlorotrimethylsilane 131
Chlorotriphenylmethane 129
Chromium(II) chloride 61
Chromosomal aberration 99
CI-920 14
Cinnamic acid 27, 28
Circular dichroism 1
β-Citromycinone 57
Claisen condensation 59
retro-Claisen reaction 49
Claisen rearrangement 46, 56
^{13}C-NMR spectroscopy 20
– 6-Substituted-5,6-dihydro-α-pyrones 20
Componotus sp. 29
Compositae 5, 6, 14, 28
Conifers 171
Cori ester 126
Cotton effect 7, 12, 19, 23, 24, 25
Cotton seed meal 170
p-Coumaric acid 27, 28
Cruciferae 28
Cryopreservation 99
Cryptocarya bourdilloni 18
Cryptocarya caloneura 17
Cryptocaryalactone 18, 24, 28
Cryptocarya massoia 5
Cryptocarya moschata 18
Cryptocarya sp. 18
Cyanide 41, 42, 74
Cyanohydrins 74
Cyanophthalides 62, 73, 74
Cyclohexylidene dimethyl acetal 133

Daunomycinone derivatives 43
Daunomycinones 49, 50, 52, 66, 67, 72
Daunorubicin 38
5-Deacetoxy-5'-epi-olguine 12, 13
Deacetylboronolide 6
Deacetylumuravumbolide 11
(+)-5-Decanolide 10
(+)-4-Demethoxydaunomycinone 66
4-Demethoxy-7-deoxy-daunomycinone 39
6-Deoxy-L-allose 13
11-Deoxyanthracyclines 67
(−)-7-Deoxydaunomycinone 60
6-Deoxydaunomycinone 64
11-Deoxydaunomycinones 50, 56, 57, 65, 67
(−)-1-Deoxyerythritol 9
(−)-(2*R*,3*S*)-1-Deoxyerythritol 9
1-Deoxy-1-fluoro-D-fructose 126
4-Deoxy-4-fluoro-galactosucrose 157
6-Deoxy-6-fluoro-galactosucrose 157
1'-Deoxy-1'-fluorosucrose 126
6-Deoxy-6-fluorosucrose 157
1-Deoxy-β-D-fructofuranosyl α-D-glucopyranoside 126
4-Deoxy-β-D-fructofuranosyl α-D-glucopyranoside 126
6-Deoxy-β-D-fructofuranosyl α-D-glucopyranoside 126
2-Deoxy-α-D-glucopyranosyl β-D-fructofuranoside 126
3-Deoxy-α-D-glucopyranosyl β-D-fructofuranoside 126
4-Deoxy-α-D-glucopyranosyl β-D-fructofuranoside 126
6-Deoxy-α-D-glucopyranosyl β-D-fructofuranoside 126
4-Deoxy-γ-rhodomycinone 67
4-Deoxysucrose 157, 166
6-Deoxysucrose 158, 166
6-Deoxy-β-D-xylo-hex-5-enopyranosyl 6-deoxy-β-D-threo-hex-5-enofuranosides 157
Desacetylakuammiline 91, 93
Dextran 157
Dextro-rotatory 10,11-dihydroxydodecanoic acid 12
Diacetone glucose 68
3,4-Di-*O*-acetyl-L-rhamnal 4
1',6'-Diamino-1',6'-dideoxysucrose 160
6,6'-Diamino-6,6'-dideoxysucrose 159
1',4':3',6'-Dianhydrosucrose 151
3,6:3',6'-Dianhydrosucrose 149, 150, 151

Diazomethane-borontrifluoride etherate 130
DIBAH reduction 61, 69
6,6′-Dibromo-6,6′-dideoxysucrose hexaacetate 163
α,α-Dibromotoluene 132
6,6′-Di-O-t-butyldiphenylsilyl sucrose 131
4,6-Dichloro-4,6-dideoxy-α-D-galactopyranosyl 3′,4′-anhydro-1′,6′-dichloro-1′,6′-dideoxy-β-D-ribohexulofuranoside 2,3-sulphate 151
4,6-Dichloro-4,6-dideoxy-galactosucrose 156
6,6′-Dichloro-6,6′-dideoxysucrose 133, 156
6,6′-Dichloro-6,6′-dideoxysucrose 3,4,3′,4′-tetra-acetate 148
Dichloromethane 130
2,4-Dichlorophenoxyacetic acid 102
Dicyclohexylcarbodiimide 74
N,N-Dicyclohexylcarbodiimide 139
1,2-Dideacetylboronolide 6
7,9-Dideoxydaunomycinone 41
4,6-Dideoxy-4,6-difluoro-galactosucrose 157
6,6-Dideoxysucrose 164
6,6′-Dideoxysucrose 126, 157
Diels-Alder reaction 42, 43, 44, 49, 58, 59, 61, 63, 64, 65, 72, 75, 77
Diethyl azodicarboxylate 152
Diethyl carbonate 149
(+)-(R,R)-Diethyl tartrate 7
Dihydrokawain 18, 19, 24
Dihydrokawain-5-ol 19, 24
(+)-Dihydrokawain-5-ol 19
Dihydromethysticin 18, 19, 24
Dihydronaphthalene 41, 59
Dihydroosmundalactone 8
5,6-Dihydro-2H-pyran-2-one 2
5,6-Dihydro-α-pyrones 1, 2, 3, 8, 9
Dihydrositsirikine 91, 93
1,4-Dihydroxyanthrahydroquinones 67
1,4-Dihydroxy-9,10-anthraquinones 70
2,1′:4,6-Di-O-isopropylidene derivative 132
1,2:4,6-Di-O-isopropylidene-α-D-glucopyranose 133
2,1′:4,6-Di-O-isopropylidene sucrose 143
4,6:1′,2-Di-O-isopropylidene sucrose 137, 138, 175
4,6:2,1′-Di-O-isopropylidene sucrose 152

2,1′:4,6-Di-O-isopropylidenesucrose tetraacetate 137
1′,6′-Di-O-mesyl-sucrose 151
1,4-Dimethoxy-benzaldehyde 49
(−)-(2S,3S)-1,4-Dimethoxy-2,3-butanediol 60
3″,4″-Dimethoxydihydrokawain 24
4″,3″-Dimethoxydihydrokawain 19
Dimethoxyphenylsilane 134
2,2-Dimethoxypropane 132, 133
1′,6′-Di-O-methyl sucrose 130
4,6-Di-O-methyl sucrose 130
6,6′-Di-O-methyl sucrose 130
4-Dimethylaminopyridine 131
N,N-Dimethyldithiocarbamate 163
N,N-Dimethylformamide 132, 134, 138, 143, 148, 149, 152, 153, 155, 156, 159, 164, 175
Dimethyl (3-oxopentyl)malonate 63
2,2-Dimethylpropionyl chloride 139
Dimethyl sulphate 130, 131
Dimethyl sulphoxide 158
N,N-Dimethyltryptamine 91, 95
6,6′-Di-O-palmitoylsucrose 130
6,6′-Dithiosucrose 163
1′,3′-Ditosylate 154
1′,4′-Ditosylate 154
6,6′-Di-O-tosyl-sucrose 151
6,6′-Di-O-tosylsucrose hexa-acetate 154
6,6′-Di-O-tosylsucrose hexabenzoate 154, 159
6,6′-Di-tritylsucrose 130
1′,6′-Di-O-tritylsucrose 129
6,1′-Di-O-tritylsucrose 129
6,6′-Di-O-tritylsucrose 129, 142
1′,6′-Di-O-tritylsucrose hexa-acetate 130
6,6′-Ditritylsucrose hexabenzoate 135
Douglas fir 170
Doxorubicin 38
Drierite 170

Elbs oxidation 75
Enzyme-linked immunosorbent assay 97, 98
19-Epiajmalicine 96
6,6′-Epidisulphide 163
(−)-Epipestalotin 7
6,6′-Epithiosucrose hexa-acetate 164
19-Epivindolinine 91, 93
19-Epivindolinine-N_b-oxide 91, 93

Subject Index

(−)-2,3-Epoxybutyric acid 9
(−)-(2R,3S)-Epoxybutyric acid 9
Erlose 169
Escherichia coli 171
Ethanol 149, 164
Ethyl chloroformate 149
6-Ethyl-5,6-dihydro-2H-pyran-2-one 2
Ethylene dichloroformate 149
Ethyl evernate 75
Ethylmagnesium bromide 68
5-Ethyl-2-penten-5-olide 2
Ethynyl magnesium bromide 59, 68
Eupatorium pilosum 6
Eurema hecabe mandarina 29

Fetizon oxidation 62
Fischer cyclisation 161
Fluorescence assay 98
Formaldehyde 58
Formic acid 55, 119
Formyl-N,N-diethylbenzamides 74
Friedel-Crafts reaction 39, 40, 42, 50, 73, 77
Fries rearrangements 56
β-D-Fructofuranosidase 121, 127, 155
β-D-Fructofuranosyl 6-deoxy-α-D-xylo-hex-5-enopyranoside 158
β-D-Fructofuranosyl derivatives 168
β-D-Fructofuranosyl α-D-glucopyranoside 121
1′-O-(β-D-Fructofuranosyl)sucrose 169
6-O-(β-D-Fructofuranosyl)sucrose 169
6′-O-(β-D-Fructofuranosyl)sucrose 169
α-D-Fructofuranosyl 1-thio-α-D-glucopyranoside 125
Fructo-oligosaccharides 169
Fructose 101, 126, 171
D-Fructose 119, 121, 126, 132, 133
L-Fructose 123, 166
Fructose 6-phosphate 126
Fungal metabolites 7
α-D-Galactopyranosyl derivatives 170, 171
β-D-Galactopyranosyl derivatives 170, 171
α-D-Galactopyranosyl β-D-fructofuranoside 122
6-O-(O-α-D-Galactopyranosyl 1 → 6-α-D-galactopyranosyl)sucrose 171
2-O-(α-D-Galactopyranosyl)sucrose 171
6-O-(α-D-Galactopyranosyl)sucrose 166, 170
6′-O-(α-D-Galactopyranosyl)sucrose 171

4-O-(β-D-Galactopyranosyl)sucrose 171
4′-O-(β-D-Galactopyranosyl)sucrose 171
6-O-(β-D-Galactopyranosyl)sucrose 171
α-D-Galactoside galactohydrolase 170
Galactosucrose 122, 128, 139, 145, 146, 152, 155, 156, 157, 166, 167
Galactosylsucroses 171
Gallotannins 171
Galloyl derivatives 171
Gas chromatography 97
Gastric ulcers 149
Gentian 170
Geranylnerol 14
Gibberellic acid 28
Gibberellin A_3 28
Glucofuranosyl fructofuranoside 119, 120
α-D-Glucopyranosidase 127
α-D-Glucopyranosyl 4-acetamido-4-deoxy-β-D-glucoheptulopyranoside 162
α-D-Glucopyranosyl 4-deoxy-4-C-methyl-4-nitro-α-L-glucoheptulopyranoside 162
α-D-Glucopyranosyl 6-deoxy-β-D-threo-hex-5-enofuranoside 158
α-D-Glucopyranosyl derivatives 169
α-D-Glucopyranosyl 3,4-epoxy-β-D-lyxo-hexulofuranoside 152
Glucopyranosyl fructofuranoside 119, 120
α-D-Glucopyranosyl β-D-fructofuranoside 120, 121
β-D-Glucopyranosyl α-D-fructofuranoside 122
Glucopyranosyl fructo-oxetanoside 119, 120
α-D-Glucopyranosyl α-L-sorbofuranoside 126
3′-O-(α-D-Glucopyranosyl)sucrose 170
4-O-(α-D-Glucopyranosyl)sucrose 169
6-O-(α-D-Glucopyranosyl)sucrose 170
α-D-Glucopyranosyl β-D-threopentuloside 126
Glucose 67, 101, 118, 171
D-Glucose 7, 13, 16, 26, 119, 121, 132, 133, 168
[1-^{14}C]-D-Glucose 26
[2-^{14}C]-D-Glucose 26
[6-^{14}C]-D-Glucose 26
L-Glucose 123, 166
Glucoseptanosyl fructofuranoside 119, 120
α-Glucosidase 127
α-D-Glucosidase 121
α-D-Glucosyl 1-phosphate 126

Subject Index

Glutathione 28
(+)-(R)-Glyceraldehyde 5
Glycosides 171
Glycosidic anthracyclines 37
Glycosyl glycoside 121
Goniodiol 17
Goniodiol diacetate 17
Goniodiol monoacetate 17
Goniothalamin 16, 17, 24, 25, 28, 29
(+)-(R)-Goniothalamin 17
Goniothalamus andersonii 17
Goniothalamus grifithii 17
Goniothalamus macrophyllus 17, 18, 29
Goniothalamus malayanus 17
Goniothalamus sesquipedalis 17
Goniothalamus sp. 17
Goniotriol 17
Grignard addition 60
Grignard reaction 41, 47, 61
D-Gulonolactone 13

Hayashi rearrangement 40, 41
2,3,4,1',3',4',6'-Hepta-O-acetylsucrose 170
2,3,4,6,1',3',4'-Hepta-O-acetylsucrose 136, 137
2,3,4,6,1',3',6'-Hepta-O-acetylsucrose 136, 137, 172
2,3,6,1',3',4',6'-Hepta-O-acetylsucrose 136, 137
(5'Z, 8'Z, 11'Z)-6-(Heptadeca-5',8',11'-trienyl-1'-yl)-5,6-dihydro-2H-pyran-2-one 13
2-Hepten-5-olide 2
6-Heptyl-5,6-dihydro-α-pyrone 5
2,3,1',3',4',6'-Hexa-O-acetylsucrose 156
2,3,4,1',3',4'-Hexa-O-benzoylsucrose 139
Hexamethyldisilazane 131
Hexamethylphosphoric triamide 147, 154, 155, 159, 160
Hexanal 14
(+)-(S)-Hexan-1,5-diol 4
Hexane 131
1,4,5-Hexanetriol 8
High pressure liquid chromatography 97
^{1}H-NMR spectroscopy 20
- 6-Substituted-5,6-dihydro-α-pyrones 20
Homophthalic anhydride 60
Horeau method 7
Hörhammericine 91, 94
Hörhammerinine 91, 94
Hudson's lactone rule 5
Hydrazine 139

Hydroanthraquinones 56
Hydrogen chloride 49
Hydrogen cyanide 59, 74
Hydrogen fluoride 39, 41, 49
Hydrogen halide 157
Hydroiodic acid 14
Hydroxyanthraquinones 46
p-Hydroxybenzoic acid 16
4-Hydroxybutanal 8
α-Hydroxy butyric acid 69
(+)-(3S)-Hydroxybutyric acid 4
21-Hydroxycyclolochnerine 91, 93
10-Hydroxydesacetylakuammiline 91, 93
1-Hydroxy-4,5-dimethoxy-9,10-anthraquinone 45, 46
5-Hydroxydodecanoic acid 10
4-Hydroxy-hepta-O-pivaloylsucrose 139
7-Hydroxyindolenineajmalicine 91, 92
4''-Hydroxy-3''-methoxydihydrokawain 19, 24
19-Hydroxy-11-methoxytabersonine 91, 94
(−)-(E,R)-3-Hydroxy-5-phenyl-4-pentenoic acid 17
(+)-(E,S)-3-Hydroxy-5-phenyl-4-pentenoic acid 17
Hydroxyphthalides 73, 74, 75
3''-Hydroxypsilotin 16
19-Hydroxytabersonine 91, 94
Hyptis oblongifolia 12
Hyptis pectinata 11
Hyptis sp. 12
Hyptolide 11, 12, 20, 24, 25

Iboza 6
Ichthyothere ulei 14
Ichthyouleolide 14, 15
Idarubicins 38
Indole-3-acetic acid 102
Indole alkaloids 90
Insect anti-feedant activity 8
Insect antifeedants 2
Invertase 121, 125, 126, 127
Invert sugar 119
2-Iodo-1,4-dimethoxybenzene 51
IR spectroscopy 23
- 6-Substituted 5,6-dihydro-α-pyrones 23
3-Isoajmalicine 91, 92
3-Iso-19-epiajmalicine 91, 92
4,6-O-Isopropylidene derivative 132
Isoraffinose 171
Isosaccharino-1,4-lactone 67

α-D-Isosaccharino-1,4-lactone 51
Isositsirikine (16R, 19E) 91, 92
Isositsirikine (16R, 19Z) 91, 93
17-Isovallesiachotamine 91, 95
Isovincoside 96

Jerusalem artichoke 169
Juglone 44

Kava 18
Kawa 18
Kawain 18, 19, 24, 28
Kawain-5-ol 19
Kawa lactones 18, 29
Kazusamycin B 14
Kende ketone 76
1-Kestose 169
6-Kestose 169

Labiateae 171
Lachnellula fuscosanguinea 9
Lachnelluloic acid 9, 10, 23, 28
Lactic acid 46, 69
(S)-Lactic acid 69
C_{26} Lactone 6, 24, 25
Lactose 129, 171
Lactosylfructoside 171
Lamiaceae 3, 5, 11
Lanthanide shift reagents 20
Lauraceae 4, 5, 19
Lead tetra-acetate 161
Leuconostoc mesenteroides 157, 171
Leucoquinizarin 67, 68, 69
Levansucrase 125, 171
Lipophilicity 166, 167
Lithiodimethoxybenzene 51
2-Lithio-2-methyl-1,3-dithian 48, 59
Lithio methyl vinyl ether 59
Lithium aluminum hydride 8, 59, 62
Lithium bromide 156
Lithium chloride 155
Lithium dimethylcuprate 67, 158
LL-P880α 7
LL-P880β 7, 8
LL-P880γ 7, 8
Lochnericine 91, 94
Lochnerinine 91, 94
Loganin 104

Malic acid 51
D-Malic acid 5, 19
L-Malic acid 17

Maltose 128, 168
Malvaceae 171
Mandelic acid 60
Manganese(II) chloride 59
L-Mannitol 123
Marschalk reaction 45, 46, 47, 67, 68, 69
Massoilactone 5, 24, 29
Mass spectrometry 21, 22
– Osmundalactone 22
– Parasorbic acid 22
– 6-Substituted 5,6-dihydro-α-pyrones 21
Meerwein-Ponndorf-Verley reduction 58
Melibiose 170
Melizitose 170
Melizitose monohydrate 170
Mercuric cyanide 170
Methanesulphonyl chloride 156
Methanol 40, 71, 119, 137, 151, 160, 175
Methanolic sodium methoxide 150, 158, 161
4-Methoxybenzaldehyde dimethylacetal 175
2-Methoxybutadiene 59
1-Methoxy-1,3-butadiene 72
1-Methoxy-1,3-cyclohexadiene 72
1-Methoxy-daunomycinone 73
2-Methoxyethanol 164
3-Methoxyphthalic anhydride 41
Methylamine 161
(–)-α-Methylbenzyl amine 53
Methyl bromoacetate 149
Methyl copper 58
Methyl α,β-dichlorocrotonate 63
Methyl 4,6-dichloro-4,6-dideoxy-α-D-galactopyranoside 145
Methyl 4,6-dichloro-4,6-dideoxy-α-D-galactopyranoside 2,3-bis(chlorosulphate) 145
6-Methyl-5,6-dihydro-2H-pyran-2-one 4
(–)-N-Methyl ephedrine 62
Methyl D-fructofuranoside 121
Methyl α-D-fructofuranoside 132
Methyl β-D-fructofuranoside 166
Methyl α-D-glucopyranoside 144, 145, 166
Methyl iodide-silver oxide-acetone 130
Methyl 1,3-O-isopropylidene-α-D-fructofuranoside 133
Methylmagnesium iodide 56, 59, 158
8-Methyl-1-octadecanol 14
(S)-3-Methylpentanoic acid 175
6-O-Methylsucrose 166
Methyl sulphoxide-acetic anhydride 158

Methyl vinyl ketone 61
Methysticin 18, 19, 24
(+)-Methysticin 19
Michael addition 70, 74, 158
Michael reaction 41, 47
Minovincinine 91, 94
Mitotic activity 100
Mitotic index 100
Mitraphylline 91, 92
Molasses 5
Mono-O-acetylsucrose 135
3′,6′-Mono-anhydro-sucrose 151
Monocotyledons 169, 171
1′-O-Monogalloyl sucrose 172
2-O-Monogalloyl sucrose 172
4′-O-Monogalloyl sucrose 172
6-O-Monogalloyl sucrose 172
6′-O-Monogalloyl sucrose 172
Monohydroxyanthrahydroquinones 67
Mono-methoxymethyl ether 55
1′-Mono-O-tosylsucrose 142
Mountain ash berries 4
Myrothecium verrucia 28

Naphthacenequinones 71
Naphthalene derivatives 49
Naphthazarin 43, 65
Naphthazarin derivative 43, 44
Naphthoquinones 43, 44
Naphthylacetic acid 102
N-(R)-1-Naphthylethyl-O-(R)-2-(1-methoxycarbonyl)-pentyl carbamate 10
Neosugar 169
Nicotiana tabacum 5
Nigrospora sp. 8
Nigrospora sp. Z 1276 8
6′-O-(p-Nitrobenzene sulphonyl) sucrose 142, 151
3-Nitro-3-deoxy-α-D-glucopyranosyl 4-nitro-4-deoxy-β-D-heptulopyranoside 162, 163
Nitroethane 161, 162
Nitrogen 102
Nitromethane 161
O-Nitrophenyl β-D-galactopyranoside 171
Nitrous acid 69
Nogalamycin 49
Nogalamycinone 71
Nystose 169

Oat stalks 169
Octa-O-benzylsucrose 123
Octa-O-mesylsucrose 154, 159
Octa-O-(trimethylsilyl) sucrose 131
Olguine 12, 13, 24
Oligosaccharides 170
Onion 169
ORD data 24
- 6-Substituted 5,6-dihydro-α-pyrones 24
Osmium tetroxide 54
Osmunda japonica 8, 29
Osmundalactone 8, 22, 24, 29
- Mass spectrometry 22
Osmundalin 8
Osmunda regalis var. *spectabilis* 8
Overhauser effect 21
Oxidation 4
6-[1′-Oxo-pentyl]-5,6-dihydro-α-pyrone 7, 8
Oxygen 105
Ozonolysis 4, 9

Paracotin 3
Parasorbic acid 4, 5, 21, 22, 23, 24, 26, 28
- Mass spectrometry 22
Parasorbiside 4
PD 113270 14
PD 113271 14
Penicillium sp. 2, 7
2,3,4,3′,4′-Penta-O-acetylsucrose 135, 136, 142
2,3,6,3′,4′-Penta-O-acetylsucrose 130, 135, 136
2,3,4,3′,4′-Penta-O-benzoylsucrose 139
2,3,6,1′,6′-Penta-O-benzoylsucrose 139, 152
Periodate 119
Periodate oxidation 4, 9
Pestalotia cryptomeriaecola 7
Pestalotin 7, 24, 28
(−)-Pestalotin 7
Pfitzner-Moffatt reagent 158
Phacelocarpus labillardieri 14
Pharmacological activity 29, 90
Pharmacology 2
Phenylalanine 27, 28
(S)-[1′-^{14}C]-Phenylalanine 28
(RS)-[2′,3′-^{13}C$_2$]-Phenylalanine 28
Phenylboronic acid 52, 54
Phenylcoumalin 3
Phenylsilylene derivatives 134
Phomalactone 8, 9, 24, 25
Phoma minispora 8

Phosphorous ylide 59
Phosphorus 102
Phosphorus acid 14
Phthalic acid derivatives 73
Piperaceae 4, 18
Piper methysticum 2, 18, 19, 29
Piperolides 18
Pirkle's method 10
Pivalates 139
Pivalic aldehyde 51
Pivaloyl chloride 139, 140
Plantaginaceae 171
Planteose 171
Plant growth inhibitors 1, 2
Plaque bacteria 157
Pleiocarpamine 91, 93
Podoblastin A 10, 28
Podoblastin B 10, 28
Podoblastin C 10, 28
Podophyllum peltatum 10
Polianthes tuberosa 10
Potassium carbonate 71, 136, 154
Potassium chloride 155
Potassium dihydrogen phosphate 102
Potassium *O*-ethyldithiocarbonate 164
Potassium hexafluorophosphine 156
Potassium hydroxide 157
Potassium iodide 55
Potassium permanganate 65
Potassium selectride 53
Potassium sucrose octasulphate heptahydrate 149
Potassium thioacetate 164
Potassium trithiocarbonate 164
6-Propyl-5,6-dihydro-α-pyrone 5
Pseudoindoxylajmalicine 91, 92
Pseudomonas saccharophila 126
Psilotaceae 16
Psilotin 16, 23, 24, 26, 27, 28
– Biosynthesis 27
Psilotin epoxide 27, 28
Psilotinin 16, 27, 28
Psilotum nudum 16
Pyran-2-one 2
Pyridine 129, 130, 131, 132, 134, 135, 138, 139, 141, 144, 146, 149, 154, 156, 157, 160, 172
Pyridine-chloroform 138
Pyridinium chloride 147, 148
α-Pyrones 1, 2, 3
2-Pyrones 2
Pyrromycinone 71

Quinine 167, 168
ortho-Quinodimethanes 61

Radioimmunoassay 97, 98
Raffinose 166, 170, 171
Raffinose pentahydrate 170
Raney nickel 157, 158, 162, 164
Reserpine 90
L-Rhamnose 8, 26
β-Rhodomycinone 70
γ-Rhodomycinone 45, 46
ε-Rhodomycinone 71
ζ-Rhodomycinone 46, 47
Rhodomycinones 42, 43, 45, 46, 67
Rhubarb 172
Rosaceae 4
Ruthenium tetroxide 58

Saccharides 169
Saccharin 166
Saccharum officinarum 119
Schiff bases 49
Secologanin 96, 103, 104, 107
Selenium dioxide 19
Serpentine 90, 91, 92, 98, 104
Sharpless epoxidation 58, 62
Silica gel 156
Silica gel chromatography 11
Silver fluoride 157
Sitsirikine 91, 92
Snatzke's rules 23, 24, 25, 26
Sodium [1-^{14}C]-acetate 26
Sodium [2-^{14}C]-acetate 26
Sodium azide 159, 160
Sodium benzoate 143
Sodium borohydride 161
Sodium bromide 155
Sodium chloride 147, 154, 155, 156
Sodium dihydrogen phosphate 102
Sodium dimethylsulfoxide 52
Sodium hydride 138
Sodium hydride-methyl iodide-*N*,*N*-dimethylformamide 130
Sodium hydrogen carbonate 149
Sodium hydroxide 131, 149
Sodium iodate 61
Sodium iodide 154
Sodium [1-^{14}C]-malonate 26
Sodium metabisulphite 163
Sodium metaperiodate 164
Sodium methoxide 151, 154, 157, 175

Sodium periodate 58
Solanaceae 5, 28, 171
Solanum berthaultii 173
D-Sorbitol 26
L-Sorbose 126
Sorbus aucuparia 4, 26
Sphaerococaeceae 14
Stachybotrys atra 28
Stachyose 171
Streptococcus mutans 157
Streptomyces pulveraceus 14
Streptomyces sp. 37
Streptomyces sp. 81–484 14
Strictosidine 91, 92, 96
Strictosidine lactam 91, 95
6-Substituted 5,6-dihydro-α-pyrones 1, 3, 11, 20, 21, 23, 24, 25, 26, 28
– Biosynthesis 26
– CD data 24
– ^{13}C-NMR spectroscopy 20
– ^{1}H-NMR spectroscopy 20
– IR spectroscopy 23
– Mass spectrometry 21
– ORD data 24
– UV spectroscopy 23
B Subtilicin 148
Sucralfate 149
Sucralose 167, 168
Sucrose 101, 102, 118, 119, 120, 121, 122, 123, 124, 125, 126, 127, 128, 129, 130, 131, 132, 133, 134, 135, 138, 139, 140, 141, 142, 143, 145, 146, 147, 148, 149, 150, 151, 152, 153, 154, 155, 156, 157, 158, 159, 160, 161, 162, 163, 164, 165, 166, 167, 168, 169, 170, 171, 172
Sucrose A 119
Sucrose B 119
C-Sucrose 124, 125
D-Sucrose 123
iso-Sucrose 122
L-Sucrose 123, 166
S-Sucrose 124, 125, 126
Sucrose 1′-aldehyde 159
Sucrose aluminium sulphate 149
4′-*O*-Sucrose 2-*O*-L-arabinopyranosyl phosphate 172
Sucrose 1′-butyrate 148
Sucrose 2,3-epoxide 164
Sucrose epoxides 153
Sucrose 2,3,6,1′,3′,4′,6′-heptapivalate 142
Sucrose monoacetate 135
Sucrose octa-acetate 122, 134, 136, 154

Sucrose octa(chlorosulphate) 146
Sucrose octamesylate 155
Sucrose octapivalate 140
Sucrose 6′-phosphate 126
Sucrose sulphonates 140
Sulfonylphthalides 73, 74
Sulfuric acid 40
Sulphuryl chloride 143, 144, 146, 151
Sulphuryl chloride-pyridine-chloroform 147
Syncolostemon rotundifolius 13
Synrotolide 13, 20, 24

Tabersonine 91, 94
Tarchonatus lactone 5
Tarchonatus trilobus 5
Tartaric acid 63
1,3,4,6-Tetra-*O*-acetyl-D-fructofuranose 122
1,3,4,6-Tetra-*O*-acetyl-α,β-D-fructofuranose 123, 124
1,3,4,6-Tetra-*O*-acetyl-D-fructofuranosyl chloride 122
2,3,4,6-Tetra-*O*-acetyl-α-D-glucopyranosyl derivatives 122
3,3′,4′,6′-Tetra-*O*-acetylsucrose 130, 143
2,3,4,6-Tetra-*O*-acetyl-1-thio-α-D-glucose 125
Tetra-aldehyde 119
1,3,4,6-Tetra-*O*-benzoyl-D-fructofuranose 123
1,3,4,6-Tetra-*O*-benzyl-D-fructofuranose 123, 125
Tetra-*O*-benzyl-α-D-galactopyranosyl bromide 170
2,3,4,6-Tetra-*O*-benzyl-D-glucopyranoside 124, 125
2,3,4,6-Tetra-*O*-benzyl-α-D-glucopyranosyl chloride 123
Tetrabutyl ammonium fluoride 157
Tetrabutylbutylammonium fluoride 131
Tetrachlorosucrose 157
4,6:4′,6′-Tetrachloro-4,6:4′,6′-tetradeoxy-galactotrehalose 168
2,6,1′,6′-Tetrachloro-2,6,1′,6′-tetradeoxy-mannosucrose 148
Tetracyclines 37
Tetradenia barberae 5
Tetradenia fruticosa 5
Tetradenia riparia 6, 11
Tetrahydroalstonine 91, 92, 96
Tetrahydrophthalic ester 64

(+)-5,6,1',2'-Tetrahydroyangoin 19
β-Tetralone derivatives 59
α-Tetralones 49, 54, 56, 57
β-Tetralones 49
1,3,4,6-Tetra-O-methyl-D-fructose 119
2,3,4,6-Tetra-O-methyl-D-glucose 119
Tetramethylidene-oxabicyclo[2.2.1]heptane 66
2,6,1',6'-Tetra-O-tosylsucrose 150
Thiele-Winter reaction 75
Thionyl chloride 175
Thiourea 163
D-Threopentulose 126
Tin tetrachloride 51, 52
Tmesipteris tannensis 16
Tobacco 173, 174
Toluene 139
Toluene-p-sulphonic acid 132, 175
Toluene-p-sulphonyl chloride 141, 154
Tosyl chloride 141, 142
Tosylhydrazone 64
Toxin 1 15, 16, 24, 28
Transition metals 48
Trehalose 168
6,1',6'-Triacetamido-6,1',6'-trideoxysucrose penta-acetate 160
3,4,6-Tri-O-acetyl-1,2-anhydro-α-D-glucopyranose 122
4,1',6'-Triamino-4,1',6'-trideoxy-galactosucrose 159, 160
3,6:1',4':3',6'-Trianhydride 150
3,6:1',4':3',6'-Trianhydro-galactosucrose 151
3,6:1',2:3,6-Trianhydrosucrose 150, 151
3,6:1',4':3',6'-Trianhydro-2-O-tosylsucrose 151
4,6,6'-Triazido-4,6,6'-trideoxy-galactosucrose 159
1,2,4-Triazole 172
2,1',6'-Tri-O-benzoyl sucrose 139
6,1',3'-Tri-O-benzoyl sucrose 139
4,6,6'-Tribromo-4,6,6'-trideoxy-galactosucrose pentamesylate 156
6,1',6'-Tri-O-t-butyldiphenylsilyl sucrose 131
Tri-n-butyltin hydride 157
Trichloroacetyl isocyanate 140
Trichloroethylbutyrate 148
2,2,2-Trichloroethylphosphorodichloridate 172
4,1',6'-Trichloro-4,1',6'-trideoxy-galactosucrose 147, 167

4,6,6'-Trichloro-4,6,6'-trideoxy-galactosucrose penta(chlorosulphate) 146
6,1',6'-Trichloro-6,1',6'-trideoxysucrose 156
6,8-Tridecanedione 10
6,1',6'-Trideoxysucrose 157
Triethylsilane 49, 54
Trifluoroacetic acid 42, 76
Trifluoroacetic anhydride 39, 42, 158
1,4,5-Trihydroxy-anthraquinone 43, 69
8,10,11-Trihydroxydodecanoic acid 12
2,4,6-Tri-isopropylbenzenesulphonyl chloride 142
1,2,4-Trimethoxybenzene 75
2,4,6-Trimethylbenzenesulphonyl chloride 142
Trimethylchlorosilane 63
Trimethylorthoacetate 175
2-Trimethylsiloxy-1,3-butadiene 60
Trimethylsilyl cyanide 74
Triphenylphosphine 14, 152, 156
Triphenylphosphine dihalide 156
Triphenylphosphine-N-halosuccinimide 156
Trisaccharides 169
Tris(dimethylamino)phosphine 156
Tris(triphenylphosphine)chlororhodium 75
Tris(triphenyl)rhodium (I) chloride 175
1',3',4'-Tritosylate 154
4,1',6'-Tri-O-tosylsucrose 151
6,1',6'-Tri-O-tosylsucrose penta-acetate 154
6,1',6'-Tri-O-trimsylsucrose 155
6,1',6'-Tri-O-trimsyl-2-O-tosylsucrose 150
6,1',6'-Tri-O-tripsylsucrose 160
6,1',6'-Tri-O-tripsylsucrose pentabenzoate 154
6,1',6'-Tri-O-tritylsucrose 129, 130, 142, 152
6,1',6'-Tri-O-tritylsucrose penta-acetate 129, 135, 136
Tritylchloride 130, 142
Tritylether 135
1'-O-Tritylsucrose 129, 130
6-O-Tritylsucrose 129
6'-O-Tritylsucrose 129, 130
6'-O-Tritylsucrose hepta-acetate 142
Tryptamine 96, 103, 104, 107
Tryptophan 103, 104
Tryptophan decarboxylase activity 102
Tuberolactone 10
Tuberose flowers 5

Subject Index

Umbelliferae 171
Umbelliferose 171
Umuravumbolide 11
$\alpha\beta$-Unsaturated-δ-lactones 2, 4, 23, 24, 25
Uridine diphosphate glucose 125, 126
UV spectroscopy 23
— 6-Substituted 5,6-dihydro-α-pyrones 23

Vallesiachotamine 91, 95
Vicia faba 170
Vilsmeyer reaction 56
Vinblastine 90, 91, 95, 97
Vincristine 90, 91, 95, 97
Vindoline 91, 94, 106, 107
Vindolinine 91, 93

Vindolinine-N_b-oxide 91, 93
Vinervine 91, 95
Vinyllithium 64

Wild potato species 173
Wine 5
Withanolides 2
Wittig reaction 11, 13, 159
Wittig reagent 158

D-Xylulose 126

Yanqona 18
Yohimbine 91, 93

Zinc borohydride 52, 58

Composition: Universitätsdruckerei H. Stürtz AG, D-8700 Würzburg
Printed by novographic, Ing. W. Schmid, A-1238 Wien

Fortschritte der Chemie organischer Naturstoffe
Progress in the Chemistry of Organic Natural Products

Volume 54:

1988. VII, 353 pages. Cloth DM 320,–, öS 2240,–.
ISBN 3-211-82086-8

Contents: T. Murakami and N. Tanaka: Occurrence, Structure and Taxonomic Implications of Fern Constituents.

Volume 53:

1988. 72 figures. VIII, 311 pages. Cloth DM 275,–, öS 1930,–.
ISBN 3-211-82074-4

Contents: L. F. Alves: Chemical Ecology and the Social Behavior of Animals – T. Nomura: Phenolic Compounds of the Mulberry Tree and Related Plants – A. Chimiak and M. J. Milewska: N-Hydroxyamino Acids and Their Derivatives.

Volume 52:

1987. 65 figures. VIII, 224 pages. Cloth DM 210,–, öS 1470,–.
ISBN 3-211-81989-4

Contents: U. Weiss, L. Merlini, and G. Nasini: Naturally Occurring Perylenequinones – H. Achenbach: The Pigments of the Flexirubin-Type. A Novel Class of Natural Products – T. Goto: Structure, Stability and Color Variation of Natural Anthocyanins – P. Bhattacharyya and D. P. Chakraborty: Carbazole Alkaloids.

Volume 51:

1987. VII, 317 pages. Cloth DM 280,–, öS 1960,–.
ISBN 3-211-81972-X

Contents: M. Gill and W. Steglich: Pigments of Fungi (Macromycetes).

Volume 50:

1986. 71 figures. IX, 261 pages. Cloth DM 210,—, öS 1470,—.
ISBN 3-211-81969-X

Contents: L. Jaenicke and F.-J. Marner: The Irones and Their Precursors — M. Lounasmaa and P. Somersalo: The Condylocarpine Group of Indole Alkaloids — U. Séquin: The Antibiotics of the Pluramycin Group ($4H$-Anthra[1,2-b]pyran Antibiotics) — R. M. Wenger: Cyclosporine and Analogues — Isolation and Synthesis — Mechanism of Action and Structural Requirements for Pharmacological Activity — H. Inouye and S. Uesato: Biosynthesis of Iridoids and Secoiridoids.

Volume 49:

1986. VIII, 400 pages. Cloth DM 290,—, öS 2030,—.
ISBN 3-211-81910-X

Contents: R. A. Hill: Naturally Occurring Isocoumarins — R. Wijnsma and R. Verpoorte: Anthraquinones in the Rubiaceae — H. Chr. Krebs: Recent Developments in the Field of Marine Natural Products with Emphasis on Biologically Active Compounds.

Volume 48:

1985. 33 figures. IX, 285 pages. Cloth DM 220,—, öS 1540,—.
ISBN 3-211-81886-3

Contents: P. S. Steyn and R. Vleggaar: Tremorgenic Mycotoxins — R. E. Moore: Structure of Palytoxin — P. Crews and S. Naylor: Sesterterpenes: An Emerging Group of Metabolites from Marine and Terrestrial Organisms.

All Volumes and Cumulative Index 1—20 available

Price reduction for subscribers: 10%

Special reduced price (20% reduction) for the complete Series Vols. 1—54 incl. the Cumulative Index to Vols. 1—20

Springer-Verlag Wien New York

Mölkerbastei 5, A-1011 Wien
175 Fifth Avenue, New York, NY 10010, U.S.A.
Heidelberger Platz 3, D-1000 Berlin 33
37-3, Hongo 3-chome, Bunkyo-ku, Tokyo 113, Japan

Springer-Verlag Wien New York

Heinz Falk

The Chemistry of Linear Oligopyrroles and Bile Pigments

This monograph will be helpful to the specialist or researcher as well as to the newcomer in this interdisciplinary field of linear oligopyrrole chemistry, which ranges from medicinal and biological to physical sciences.

Linear oligopyrroles and bile pigments are important as antenna pigments of photosynthesis, light sensory pigments in plants, and products of animal and human metabolism. Whereas di-, tri-, penta- and polypyrroles play no part in nature, they are useful as synthons in the synthesis of e.g. porphyrins and corrins, and even as organic conductors.

Discussion of the chemistry of linear oligopyrroles is started with reviews of nomenclature, occurrence, formation, importance, and history. Their structural and stereochemical aspects are illustrated by ball and stick models of X-ray crystallographic determinations as well as by the results obtained by various methods for their state of solution. The synthesis of these compounds is treated in a methodological way providing typical examples instead of listing all syntheses executed so far. Selected physical properties like absorption, emission, chiroptical data, and nuclear magnetic resonance are covered in detail. Nucleophilic, electrophilic, and radical reactions are discussed from the standpoint of semiempirical calculations providing typical examples. Moreover, their photochemistry, carrier mediated transport, skeletal transformations, redox properties, and catalytic function are included.

The book will provide advanced students approaching the subject from a variety of disciplines with the chemical background necessary to cope with the sometimes rather complicated material, but it will also provide the active researcher in this field with a timely review to inspire future work.

1989. 344 figures.
XI, 621 pages.
Cloth DM 270,-, öS 1890,-
ISBN 3-211-82112-0

Springer-Verlag Wien New York
Moelkerbastei 5, P.O. Box 367, A-1011 Wien
Heidelberger Platz 3, D-1000 Berlin 33
175 Fifth Avenue, New York, NY 10010, USA
37-3, Hongo 3-chome, Bunkyo-ku, Tokyo 113, Japan